Implications of Emerging Micro- and Nanotechnologies

Committee on Implications of Emerging Micro- and Nanotechnologies
Air Force Science and Technology Board
Division on Engineering and Physical Sciences
NATIONAL RESEARCH COUNCIL
OF THE NATIONAL ACADEMIES

THE NATIONAL ACADEMIES PRESS
Washington, D.C.
www.nap.edu

THE NATIONAL ACADEMIES PRESS • 500 Fifth Street, N.W. • Washington, DC 20001

NOTICE: The project that is the subject of this report was approved by the Governing Board of the National Research Council, whose members are drawn from the councils of the National Academy of Sciences, the National Academy of Engineering, and the Institute of Medicine. The members of the committee responsible for the report were chosen for their special competences and with regard for appropriate balance.

This study was supported by Contract/Grant No. F49620-01-1-0438 between the National Academy of Sciences and United States Air Force. Any opinions, findings, conclusions, or recommendations expressed in this publication are those of the author(s) and do not necessarily reflect the views of the organizations or agencies that provided support for the project.

International Standard Book Number 0-309-08623-X

Library of Congress Catalog Card Number: 2002115988

Additional copies of this report are available from the National Academies Press, 500 Fifth Street, N.W., Lockbox 285, Washington, DC 20055; (800) 624-6242 or (202) 334-3313 (in the Washington metropolitan area); Internet, http://www.nap.edu

Copyright 2002 by the National Academy of Sciences. All rights reserved.

Printed in the United States of America

THE NATIONAL ACADEMIES
Advisers to the Nation on Science, Engineering, and Medicine

The **National Academy of Sciences** is a private, nonprofit, self-perpetuating society of distinguished scholars engaged in scientific and engineering research, dedicated to the furtherance of science and technology and to their use for the general welfare. Upon the authority of the charter granted to it by the Congress in 1863, the Academy has a mandate that requires it to advise the federal government on scientific and technical matters. Dr. Bruce M. Alberts is president of the National Academy of Sciences.

The **National Academy of Engineering** was established in 1964, under the charter of the National Academy of Sciences, as a parallel organization of outstanding engineers. It is autonomous in its administration and in the selection of its members, sharing with the National Academy of Sciences the responsibility for advising the federal government. The National Academy of Engineering also sponsors engineering programs aimed at meeting national needs, encourages education and research, and recognizes the superior achievements of engineers. Dr. Wm. A. Wulf is president of the National Academy of Engineering.

The **Institute of Medicine** was established in 1970 by the National Academy of Sciences to secure the services of eminent members of appropriate professions in the examination of policy matters pertaining to the health of the public. The Institute acts under the responsibility given to the National Academy of Sciences by its congressional charter to be an adviser to the federal government and, upon its own initiative, to identify issues of medical care, research, and education. Dr. Harvey V. Fineberg is president of the Institute of Medicine.

The **National Research Council** was organized by the National Academy of Sciences in 1916 to associate the broad community of science and technology with the Academy's purposes of furthering knowledge and advising the federal government. Functioning in accordance with general policies determined by the Academy, the Council has become the principal operating agency of both the National Academy of Sciences and the National Academy of Engineering in providing services to the government, the public, and the scientific and engineering communities. The Council is administered jointly by both Academies and the Institute of Medicine. Dr. Bruce M. Alberts and Dr. Wm. A. Wulf are chair and vice chair, respectively, of the National Research Council.

www.national-academies.org

COMMITTEE ON IMPLICATIONS OF EMERGING MICRO- AND NANOTECHNOLOGIES

STEVEN R.J. BRUECK, *Chair*, University of New Mexico, Albuquerque
S. THOMAS PICRAUX, *Vice Chair*, Arizona State University, Tempe
JOHN H. BELK, The Boeing Company, St. Louis, Missouri
ROBERT J. CELOTTA, National Institute of Standards and Technology, Gaithersburg, Maryland
WILLIAM C. HOLTON, North Carolina State University, Raleigh
SIEGFRIED W. JANSON, The Aerospace Corporation, Los Angeles
WAY KUO, Texas A&M University, College Station
DAVID J. NAGEL, George Washington University, Washington, D.C.
P. ANDREW PENZ, Science Applications International Corporation, Richardson, Texas
ALBERT P. PISANO, University of California, Berkeley
ROSEMARY L. SMITH, University of California, Davis
PETER J. STANG, University of Utah, Salt Lake City
GEORGE W. SUTTON, SPARTA, Arlington, Virginia
WILLIAM M. TOLLES, Consultant, Alexandria, Virginia
ROBERT J. TREW, Virginia Polytechnic Institute and State University, Blacksburg
MARY H. YOUNG, HRL Laboratories, Malibu, California

Liaison

Air Force Science and Technology Board

ALAN H. EPSTEIN, Massachusetts Institute of Technology, Cambridge

Staff

JAMES C. GARCIA, Study Director
JAMES E. KILLIAN, Study Director
JAMES MYSKA, Research Associate
PAMELA A. LEWIS, Senior Project Assistant
LINDA D. VOSS, Technical Writer

AIR FORCE SCIENCE AND TECHNOLOGY BOARD

ROBERT A. FUHRMAN, *Chair*, Lockheed Corporation (retired), Pebble Beach, California
R. NOEL LONGUEMARE, *Vice Chair*, Private Consultant, Ellicott City, Maryland
LYNN CONWAY, University of Michigan, Ann Arbor
WILLIAM H. CRABTREE, Consultant, Cincinnati, Ohio
LAWRENCE J. DELANEY, President, CEO, and Chairman of the Board, Areté Associates, Arlington, Virginia
STEVEN D. DORFMAN, Hughes Electronics (retired), Los Angeles, California
EARL H. DOWELL, Mechanical Engineering, Duke University, Durham, North Carolina
ALAN H. EPSTEIN, Gas Turbine Lab, Massachusetts Institute of Technology, Cambridge, Massachusetts
DELORES M. ETTER, Professor, U.S. Naval Academy, Annapolis, Maryland
ALFRED B. GSCHWENDTNER, Lincoln Laboratory, Massachusetts Institute of Technology, Lexington, Massachusetts
BRADFORD W. PARKINSON, Stanford University, Stanford, California
RICHARD R. PAUL, Vice President, Strategic Development, Phantom Works, The Boeing Company, Seattle, Washington
ROBERT F. RAGGIO, Executive Vice President, Dayton Aerospace, Inc., Dayton, Ohio
ELI RESHOTKO, Professor Emeritus, Case Western Reserve University, Cleveland, Ohio
LOURDES SALAMANCA-RIBA, Professor, Materials Engineering Department, University of Maryland, College Park
EUGENE L. TATTINI, Deputy Director, Jet Propulsion Laboratory, Pasadena, California

Staff

BRUCE A. BRAUN, Director
MICHAEL A. CLARKE, Associate Director
WILLIAM E. CAMPBELL, Administrative Officer
CHRIS JONES, Financial Associate
DEANNA P. SPARGER, Senior Project Assistant
DANIEL E.J. TALMAGE, Research Associate

Preface

Biology long ago adopted the micro- and nanoscales. The machinery of genomics is based on nanoscale interactions, and mosquitoes, ants, termites, and other insects are exquisite examples of autonomous, intelligent micromachines that engage in both independent and cooperative (swarm) behavior. While mankind's deliberate use of nanotechnology goes back at least as far as the firing of Venetian glass during the Renaissance, only today are we developing the scientific base—theory, fabrication science, materials sophistication, and measurement capabilities—for a full-scale assault on nanotechnology.

Technology has been steadily moving into the micro- and nanoscale realms for some time. Fabrication technologies for integrated circuits are at the edge of the nanoscale, with gate lengths less than 100 nm in the most advanced microprocessors. Microelectromechanical systems (MEMS) devices are integrating mechanical motion (and other properties) on the microscale with electronics and generating new approaches to applications and even new industries.

The Deputy Assistant Secretary of the Air Force for Science, Technology, and Engineering requested that the Committee on Implications of Emerging Micro- and Nanotechnologies, established by the National Research Council, assess the implications of emerging micro- and nanotechnologies for the Air Force. The committee was asked to characterize the state of the art in micro- and nanotechnologies, review the adequacy of military investment strategies for micro- and nanotechnologies, and recommend research areas to accelerate the opportunities for exploiting these technologies in Air Force mission capabilities and systems.

The committee received briefings from experts in varied aspects of micro- and nanotechnologies from within and outside the Air Force. Four implications of these evolving technologies are clear: ever-increasing information capabilities, a relentless drive toward miniaturization, new materials with new functionality based on nanoscale structuring, and higher-level systems integration, with increased functionality leading ultimately to autonomous systems. Some of the challenges are as large as the opportunities, including translating the unique properties of micro- and nanostructures into macro effects and manufacturing micro- and nanomaterials and components inexpensively on a large scale.

Suffice it to say, micro- and nanotechnologies are an important area of research opportunity at a productive stage of development. The impacts, while not entirely predictable, can be characterized in general terms and will clearly be significant. The Air Force should harness the power of these technologies for its missions.

The scope of this study was daunting, covering many orders of magnitude in spatial scale and many decades of future progress. The committee is indebted to the experts, both within and outside the Air Force, who took the time to share their insights. The committee greatly appreciates the support and assistance of National Research Council staff members James Garcia, James Killian, Pamela Lewis, and James Myska and consultant Linda Voss in the development and production of this report.

Steven R.J. Brueck, *Chair*
S. Thomas Picraux, *Vice Chair*
Committee on Implications of Emerging Micro- and Nanotechnologies

Acknowledgment of Reviewers

This report has been reviewed in draft form by individuals chosen for their diverse perspectives and technical expertise, in accordance with procedures approved by the National Research Council's Report Review Committee. The purpose of this independent review is to provide candid and critical comments that will assist the institution in making its published report as sound as possible and to ensure that the report meets institutional standards for objectivity, evidence, and responsiveness to the study charge. The review comments and draft manuscript remain confidential to protect the integrity of the deliberative process. We wish to thank the following individuals for their review of this report:

Larry R. Dalton, University of Washington, Seattle
Elsa Reichmanis, Bell Laboratories, Lucent Technologies, Murray Hill, New Jersey
Lourdes Salamanca-Riba, University of Maryland, College Park
Henry I. Smith, Massachusetts Institute of Technology, Cambridge
T.S. Sudarshan, Materials Modification, Inc., Fairfax, Virginia
Richard Taylor, Hewlett-Packard Laboratories, Bristol, United Kingdom
George M. Whitesides, Harvard University, Cambridge, Massachusetts.

Although the reviewers listed above have provided many constructive comments and suggestions, they were not asked to endorse the conclusions or recommendations, nor did they see the final draft of the report before its release. The review of this report was overseen by Royce W. Murray, University of North Carolina, Chapel Hill. Appointed by the National Research Council, he was responsible for making certain than an independent examination of this report was carried out in accordance with institutional procedures and that all review comments were carefully considered. Responsibility for the final content of this report rests solely with the authoring committee and the institution.

Contents

	EXECUTIVE SUMMARY	1
1	INTRODUCTION	20

Background, 20
Statement of Task, 22
What We Mean by "Micro" and "Nano," 22
Report Organization and Methodology, 27
References, 29

2	EXPECTATIONS FOR FUTURE MICRO- AND NANOTECHNOLOGIES	30

Overview of Current Studies, 30
    International Technology Roadmap for Semiconductors, 30
    MEMS Industry Group 2001 Annual Report, 34
    The National Nanotechnology Initiative, 36
Worldwide Perspective, 38
References, 39

3	MAJOR AREAS OF OPPORTUNITY	40

Information Technology, 40
    Introduction, 40
    Computing Capabilities—Devices, 41
    Computing Capabilities—Architectures, 56
    Storage, 61

Communications, 64
Signal and Information Processing and Data Fusion, 74
Findings and Recommendations, 78
Sensors, 80
   Introduction, 80
   Discrete Versus Distributed Sensors, 81
   Projected Impact, 81
   Sensors for Chemical and Biological Agents, 89
   Self-Sensing, 91
   Distributed Sensor Systems, 95
   Findings and Recommendations, 98
Biologically Inspired Materials and Systems, 99
   Biomimetics for Improved Sensing, Communications, and
      Signal Processing, 100
   Enhanced Human Performance—The Machine as Part of the Man, 102
   Findings and Recommendations, 103
Structural Materials, 103
   Introduction, 103
   Lightweight Materials, 105
   Improved Coatings, 107
   Multifunctional Structures, 108
   Materials for MEMS, 110
   Technical Issues and Areas for Development, 111
   Findings and Recommendations, 112
Aerodynamics, Propulsion, and Power, 113
   Flight Vehicle Aerodynamics, 114
   Air-Breathing Vehicle Propulsion and Power, 116
   Launch Vehicle Propulsion, 121
   Spacecraft Propulsion, 123
   Space Power Generation, 129
   Findings and Recommendations, 131
References and Notes, 131

4 ENABLING MANUFACTURING TECHNOLOGIES   143
   Fabrication (Patterning) Approaches, 143
      Lithography and Pattern Transfer, 145
      Self-Assembly, 152
      Integration of Traditional Lithographic and Self-Assembly
         Patterning Approaches, 154
      Integration of Nanodevices with Mainstream Silicon
         Technology, 157
   Assembly, 158
      Directed Assembly, 158

DNA-Assisted Assembly, 160
Packaging, 163
Reliability and Manufacturability, 165
  New Techniques for Reliability Improvement, 166
  Manufacturing Yield and Reliability, 166
Commercialization, 167
  Identification of Products Manufactured in the New Technology, 168
  Wide Access to the Technical Details of the New Technology, 168
  Enlightened Corporate Management, 169
  Sufficient Reduction in Product Cost, 169
  Government Role in Providing Wide Access to New Technology, 169
  Effect of Manufacturing Complexity on Commercialization, 172
  Case Study: Texas Instruments and the Digital Mirror Device, 173
Findings and Recommendations, 176
References, 178

5   AIR FORCE MICRO- AND NANOTECHNOLOGY PROGRAMS
    AND OPPORTUNITIES   182
    Impacts of Micro- and Nanotechnologies on Air Force Missions, 182
    Current Investments by the Air Force in Micro- and
        Nanotechnologies, 183
      AFRL Research Portfolio in Micro- and Nanotechnologies, 184
      AFOSR Basic Research Programs in Nanotechnology, 184
    Trends in DoD and Air Force Research Funding, 190
    Air Force Investment Strategy and Challenges, 195
    Findings and Recommendations, 197
    References, 199

6   OPPORTUNITIES IN MICRO- AND NANOTECHNOLOGIES   200
    Overarching Themes, 200
      Increased Information Capabilities, 200
      Miniaturization, 201
      New Engineered Materials, 202
      Increased Autonomy and Functionality, 202
    Air Force Missions as Drivers for Micro- and Nanotechnologies, 203
    Areas of Opportunity, 204
      Space Vehicles and Systems, 205
      Weapon Systems, 211
      Air Vehicles and Systems, 212
    Finding and Recommendation, 215
    References, 215

| 7 | FINDINGS AND RECOMMENDATIONS | 216 |

Critical Findings and Recommendations, 216
    Technology, 216
    Policy, 220
Specific Findings and Recommendations, 222

APPENDIXES

A   Manufacturing, Design, and Reliability, 229
B   Committee Biographies, 232
C   Meetings and Activities, 239

Figures, Tables, and Boxes

FIGURES

1-1   Model of a MEMS safety switch, 23
1-2   Atomic force microscopic image of InAs quantum dots, 24
1-1-1 Dimensional scale, 25
1-2-1 The SNAP-1 nanosatellite, 28

2-1   Integrated circuit growth, 31
2-2   Lithography half-pitch feature size versus time, 31
2-3   Possible roadmap, 34
2-4   Worldwide government R&D spending on nanotechnology, 39

3-1   Power versus frequency for high-frequency microwave devices, 51
3-2   Yearly radiation dose in silicon, 54
3-3   Radiation environment for circular equatorial orbits, 55
3-4   Diode laser thresholds, 66
3-5   InAs quantum dashes grown on InP, 68
3-6   Optical MEMS examples, 71
3-7   RF MEMS capacitors, 73
3-8   Schematic of a situational awareness system, 75
3-9   Paradigm shifts in software, 77
3-10 Micromachined Sun sensor, 88
3-11 Boeing/Endevco pressure belt, 92

xv

3-12 A typical pressure-sensitive paint result for a wind tunnel model of a transonic transport airplane, 93
3-13 The Very Large Array, 95
3-14 Micromachined gas turbine engine, 118
3-15 Silicon turbine from the micromachined gas turbine engine, 119
3-16 Planar glass layers for a batch-producible cold gas propulsion module, 124
3-17 Spacecraft power for INTELSAT satellites, 126
3-18 Use of a momentum-exchange tether to perform an orbit transfer, 128
3-1-1 Carbon nanotube structures, 47
3-2-1 Swarm of nanosatellites, 97

4-1 Communities needed for the production, maintenance, and use of military hardware, 144
4-2 Lithography examples, 146
4-3 The sequential steps in LIGA, 148
4-4 Schematic of the structures used in LISC, 149
4-5 Integrated circuit production, 151
4-6 Cross-sectional photograph of a silicon wafer processed by deep reactive ion etching, 152
4-7 Rotapod MEMS device, 160
4-8 Principle of DNA-assisted pick and place, 162
4-9 DNA-assisted microassembly, 163
4-10 Lenslet array fabricated using hydrophobic/hydrophilic selectivity, 164
4-11 Cumulative user accounts for the MEMS exchange, 171
4-12 Cut-away of the digital mirror device structural model, 174
4-13 Photomicrograph of the digital mirror device, 174
4-1-1 Two-dimensional active pixel sensor array, 170

5-1 Air Force nanotechnology research, 186
5-2 Trends in federal R&D funding, FY 1990–2003, 191
5-3 Funding of basic research by DoD, 191
5-4 Science and technology funding levels by Service, 192
5-5 Integrated circuit sales, 194

6-1-1 DARPA/Aerospace Corp. picosatellites, 208
6-2-1 The AeroVironment Black Widow micro air vehicle, 213
6-2-2 Subsystem layout, size, and mass of the Black Widow, 214

A-1 A computerized manufacturing procedure for nanoproducts, 231

TABLES

ES-1 Recommended Air Force Roles in Micro- and Nanotechnology Research, 7
ES-2 Taxonomy of Micro- and Nanotechnology Research Areas and Their Relevance to the Air Force, 8
ES-3 Selected Mission and Platform Opportunity Areas, 14

2-1 Predictions of 2001 ITRS for Selected Parameters, 33

3-1 Approximate Radiation Hardness Levels for Semiconductor Devices, 53

4-1 Reliability Paradigm for Nanoproducts, 167
4-2 High Complexity of the Digital Mirror Device, 175

5-1 Challenges and Impact Areas, 185
5-2 Air Force Nanotechnology Research, 186
5-3 AFOSR-Managed DURINT Programs, 189
5-4 Nanotechnology MURIs in FY 2001, 189
5-5 AFOSR Technology Grants in FY 2001, 190

6-1 Selected Mission and Platform Opportunity Areas, 206

BOXES

1-1 A Matter of Scale, 25
1-2 Small Satellites: How Small Can We Go?, 28

3-1 The Ubiquitous Carbon Nanotube, 47
3-2 Emergent Behavior of Swarms of Microplatforms, 97

4-1 MOSIS, 170

5-1 Expected Impacts of Research Supported by the Air Force Nanotechnology Program, 184
5-2 Initial DoD Focus in Nanotechnology, 188
5-3 Air Force Nanotechnology Program, 188

6-1 Nano- and Picosatellites, 208
6-2 The Black Widow Micro Air Vehicle, 213

Acronyms

AFM	atomic force microscope
AFOSR	Air Force Office of Scientific Research
AFRL	Air Force Research Laboratory
AFSTB	Air Force Science and Technology Board
APD	avalanche photo diode
ASIC	application-specific integrated circuit
ASIM	application-specific integrated microinstrument
BARC	Bead Array Counter
CA	cellular automata
CAD	computer-aided design
CAM	computer-aided manufacturing
CAPP	computer-aided process planning
CBM	condition-based maintenance
CMOS	complementary metal oxide semiconductor
CNT	carbon nanotube
CONOPS	concept of operations
CPU	central processing unit
DARPA	Defense Advanced Research Projects Agency
DDR&E	Director of Defense Research and Engineering
DMD	digital mirror device
DNA	deoxyribonucleic acid

DoD	Department of Defense
DRAM	dynamic random-access memory
DSP	digital signal processor
DURINT	Defense University research in nanotechnology
DURIP	Defense University Research Instrumentation Program
ECL	emitter-coupled logic
EDAC	error detection and control
ELINT	electronic intelligence
ELO	epitaxial lateral overgrowth
EPROM	erasable programmable read-only memory
EUV	extreme ultraviolet
FEL	free-electron laser
FY	fiscal year
GEO	geosynchronous orbit
GLOW	gross liftoff weight
GMR	giant magnetoresistive
GPS	Global Positioning System
GTO	geosynchronous transfer orbit
HEMT	high-electron-mobility transistor
IC	integrated circuit
IEEE	Institute of Electrical and Electronics Engineers
IMU	inertial measurement unit
IR	infrared
IT	information technology
ITRS	International Technology Roadmap for Semiconductors
JSEP	Joint Service Electronics Program
JSTARS	Joint Surveillance Target Attack Radar System
LANL	Lawrence Livermore National Laboratory
LCE	life-cycle engineering
LEO	low Earth orbit
LIGA	Lithographie, Galvanoformung, und Abformung
LISA	lithographically induced self-assembly
LISC	lithographically induced self-construction
MAC	MEMS-based active aerodynamic flight control vehicle
MACSAT	multiple access communications satellite

MAV	micro air vehicle
MBE	molecular beam epitaxy
MCM	multichip modules
MEMS	microelectromechanical systems
MEU	multiple-event upset
MOEMS	microoptoelectromechanical system
MOSFET	metal oxide semiconductor field effect transistor
MPG	micropower generator
MRAM	magnetic random-access memory
MURI	Multidisciplinary University Research Initiative
NDR	negative differential resistance
NEMS	nanoelectromechanical system
NIL	nanoimprint lithography
nm	nanometer
NNI	National Nanotechnology Initiative
NRC	National Research Council
PHM	condition-based and prognostics health monitoring
pico	prefix for 10^{-12}
PMMA	polymethylmethacrylate
QDCA	quantum-dot cellular automata
R&D	research and development
RDT&E	research, development, testing, and evaluation
RF	radio frequency
RTD	resonant tunneling diode
S&T	science and technology
SEM	scanning electron microscope
SEU	single-event upset
Si	silicon
SIA	Semiconductor Industry Association
SOC	system-on-a-chip
SPENVIS	Space Environment Information System
SRAM	static random-access memory
SRMU	solid rocket motor unit
STTL	Shottky transistor-transistor logic
SWNT	single-wall carbon nanotube
TFSOI	thin-film silicon-on-insulator
TTL	transistor-transistor logic

UAV	unmanned air vehicle	
UCAV	unmanned combat air vehicle	
URI	University Research Initiative	
VCSEL	vertical cavity surface-emitting laser	
VLA	Very Large Array	
WDM	wavelength division multiplexing	
WLR	wafer-level reliability	

Executive Summary

The Committee on Implications of Emerging Micro- and Nanotechnologies, established by the National Research Council's (NRC's) Air Force Science and Technology Board (AFSTB), was tasked with evaluating the implications of current trends in micro- and nanotechnologies for the Air Force. As a basis for its evaluation, the committee applied rigorous technical scrutiny to claims for the potentials of these technologies, evaluated the state of the technologies today, and assessed their value relative to enduring Air Force requirements. The committee looked for trends in scientific and technical advances with the potential to change the nature of warfare and for the most effective ways for the Air Force to exploit these advances.

Predicting the progress of technology over long periods is an uncertain exercise. The temptations are to be either too conservative, acknowledging the current limitations of technology and not foreseeing the breakthroughs—in both conception and capability—that will inevitably occur, or too exuberant, brushing aside real physical limitations in an excess of futuristic zeal. Such a challenge particularly applies to nanotechnology, which is an exciting and relatively unexplored scientific and technological frontier offering many new insights and applications but at the same time giving rise to much speculation and hyperbole. From an applications perspective, microtechnologies and nanotechnologies offer a particularly powerful combination for future Air Force missions and thus deserve careful consideration.

DEFINITION OF MICRO- AND NANOTECHNOLOGIES

In undertaking this study, the committee decided not to put hard size limitations on micro- and nano- objects and technologies. It understands these concepts as relating roughly to scale but also as differing significantly in the importance of various underlying physical and chemical mechanisms. There is no hard line between micro and nano, but there are some clear differences in the way the science and technology communities approach these regimes.

The concept of microtechnology has become somewhat familiar. One hundred million transistor computer chips are in our homes, and the public has a vague concept of the manufacturing processes, having seen many pictures of clean rooms and workers in "bunny suits." Now, microtechnology is migrating from the electronics domain into a much broader range of technologies with the introduction of microelectromechanical systems (MEMS) and biological applications such as micro-reaction arrays for drug discovery.

A defining feature of the nanoscale is that the behavior of a material differs in fundamental ways from that observed at the macro- and the microscales. New physics and chemistry come into play. Dimensions, as well as composition and structure, impact material properties in nanoscale materials. At least two factors dominate this transition. The first is that nanometer dimensions approach characteristic (quantum) wave function scales of excitations in the material—electrons and holes, photons, spinwaves, and magnons, among others. The second is the very large surface-to-volume ratio of these structures, which means that no atom is very far from an interface and that interatomic forces and chemical bonds dominate. The large surface areas and unique interface and molecule–solid interactions at nanostructure surfaces are the basis of much of the enthusiasm driving research at the boundary between nano- and biotechnologies. The information stored in the genome and the exquisite selectivity of biochemical interactions based on chemical recognition and matching are examples of nanoscale properties where interfacial forces play a determining role.

Nanotechnology is likely to require an approach to fabrication fundamentally different from that of microtechnology. Whereas microscale structures are typically formed by top-down techniques such as patterning, deposition, and etching, the practical formation of structures at nanoscale dimensions will probably involve an additional component, bottom-up assembly. Self-assembly, a process whereby structures are built up from atomic or molecular-scale units into larger and increasingly complex structures, is widely used by biological systems. As our capabilities expand, some combination of top-down (lithographic) and bottom-up (including self-assembly) techniques probably will be employed for the efficient manufacturing of nanoscale systems.

One last word on definition: Not all things "nano" adhere to the usual nanometer dimensional scale, nanosatellites being a notable example. Nanosatellites have overall dimensions of many centimeters—and the name evolved out of the need for a word to designate systems that are significantly smaller, in a revolutionary way, from today's large satellite systems. However, even here the underlying capabilities for developing nanosatellites are provided by advances in micro- and nanotechnologies.

OVERARCHING THEMES EMERGING FROM MICRO- AND NANOTECHNOLOGIES

Four overarching themes emerged from the committee's study of micro- and nanotechnologies:

- increased information capabilities
- miniaturization of systems
- new materials resulting from new science at these scales, and
- increased functionality and autonomy.

These themes are a natural consequence of the advances in micro- and nanotechnologies resulting from scaling to small size. The new capabilities these advances provide will have far-reaching consequences for Air Force missions.

Increased Information Capabilities

The committee foresees a continued scaling of microelectronic, magnetic, and optical devices to smaller size and higher densities. The result will be the ability to store, process, and communicate an ever-increasing amount of information at ever-higher speeds. The current rapid increase in the ability to handle information, enshrined in Moore's law—the exponential increase in computing power—will continue at least for the next decade. The integrated effects of the continued doubling of computing power every 18 months and the even more rapid increase in information transmission rate and storage density will lead to an increase of at least two orders of magnitude in the amount of information that can be gathered and processed. Today's smart weapons will seem "mentally challenged" by the next decade. Understanding how to utilize all of this computational capability and information availability will be a significant challenge. Beyond this relatively near-term trend, the committee anticipates that emerging nanotechnologies will enable even more revolutionary long-term changes in how we obtain and use information. Exploiting these revolutionary changes will be an important and challenging task for the Air Force.

Miniaturization

The reduction in the size of systems, from computers to cell phones, is a continuing trend for electronic systems. The significance of this miniaturization goes well beyond just the smaller size and reduced weight. Batch fabrication has been a key driver in the miniaturization of microelectronics, enabling reduced cost and increased reliability and robustness through the parallel manufacturing of many integrated components. Another significant aspect is the combination of components and subsystems into fewer and fewer chips, enabling increased functionality in ever-smaller packages. These trends are evolving to include MEMS and other technologies for sensors and actuators, making it possible to miniaturize entire systems and platforms. The combination of reduced size, reduced weight, increased robustness, and reduced cost-per-unit-function has significant implications for Air Force missions, from global reach to situational awareness. Examples may include the rapid, low-cost global deployment of sensors, launch-on-demand tactical satellites, distributed sensor networks, and affordable and highly capable unmanned air vehicles.

New Engineered Materials

Advances in micro- and nanofabrication technologies are enabling the engineering of materials down to the atomic level. While design and fabrication capabilities are still primitive from an applications perspective, there is great potential for improving the properties and functionality of materials. Examples of recent advances in materials range from carbon nanotubes with great strength and novel electronic properties, to quantum dot communication lasers, to giant magnetoresistive materials for high-density magnetic memories. Theory and simulation will play an increasingly important role in guiding the development of new nanostructured materials and of systems based on such materials. By combining materials at the micro- and nanoscales to form smart composite structures, additional increases in functionality can be achieved. New materials are an underlying enabling capability. They will be used to expand the performance of electronics, sensors, communications systems, avionics, air and space frames, and propulsion systems. Theory and modeling of materials are advancing significantly, as is our understanding of the relationships between composition and structure across many spatial scales and the resulting material properties. Over time, these advances may reduce the long lead time for developing new materials, as well as help in the design of new, more functional materials for Air Force systems.

Increased Functionality and Autonomy

Advances in information density, miniaturization, and materials will enable a degree of autonomous operation for intelligent systems that cannot be fully envisioned today. Enhanced functionality and increasing autonomy, based on micro- and nanotechnologies, will have many system benefits: lower risk for humans; higher-performance systems; lower-cost platforms; and reduced communication requirements, with a correspondingly lower probability of detection. Initially, autonomy will be seen simply as an evolutionary extension of the capabilities of current systems, such as cruise missiles or unmanned combat air vehicles, providing increased accuracy and range or other performance advantages. Over the longer term, however, the dramatic increases in local information awareness and computational power will enable independent decision making and will have a dramatic impact on the conduct of warfare. Systems may also be able to power, self-repair, and reconfigure themselves to extend and expand the scope of their mission. The lowered cost and increased functionality will lead to swarms of intelligent agents, with emergent behavior that differs from that of any single entity. Integrating these advances into the Air Force concept of operations will be challenging and will raise important global political and societal issues, such as the acceptable bounds of future warfare, including specifying the roles of autonomous decision-making machines in war fighting.

DIRECTIONS OF FUTURE MICRO- AND NANOTECHNOLOGIES

Ambitious research programs are under way in all aspects of micro- and nanotechnologies. The communities involved have looked to the future and articulated their visions of progress and the capabilities that will be useful benchmarks for the Air Force in planning its research and development programs. Three recent compilations are the International Technology Roadmap for Semiconductors (ITRS),[1] the MEMS Industry Group 2001 Annual Report,[2] and the implementation plan for the National Nanotechnology Initiative (NNI).[3]

The ITRS provides a very detailed assessment of what it will take to remain on the Moore's law scaling curve for the dominant electronics technology, complementary metal oxide semiconductor (CMOS) technology, and what the roadblocks will be to achieving this scaling.

No clear path is seen for extending CMOS technology beyond the roadmap horizon of 2016, when the projected devices will be well into the nanometer region, with physical gate lengths of only 9 nm. Indeed, a transition to an alternative technology may be needed before 2016 because of physical limitations such as the onset of quantum and interface effects and cost limitations related to the increasing difficulty of scaling the current top-down manufacturing paradigm.

On the other hand, these same nanoscale phenomena offer the opportunity of developing whole new classes of devices relying on the principles of quantum physics that may provide the basis for future information processing. However, systems based on these new principles might function in a much different manner from current computing systems.

The 2001 Annual Report of the MEMS Industry Group discussed key drivers and challenges expected in MEMS for the next 20 years. The growth of MEMS technology will be driven by optical and wireless networks as well as needs in health care and biotechnology, among other areas. A number of technology barriers related to the transition of MEMS from one-off laboratory demonstrations to a low-cost, robust, manufacturing-based industry must be overcome: the development of MEMS fabrication equipment, standardized MEMS foundry processes, MEMS-specific packaging technologies, and enhanced integration with electronics.

The research community has gained a much greater appreciation for the degree to which control of the structure of matter on the nanoscale can determine the macroscopic properties of materials. Furthermore, there have been remarkable advances in the ability to manipulate matter at the atomic and molecular levels. This has led to a dramatic increase in U.S. government funding for nanotechnology research, estimated at $600 million in FY 2002. The rest of the world is spending apace, with the total annual investment of governments around the world in nanotechnology research now estimated at over $1.6 billion.

The NNI set forth eight grand challenges that provide perspective on the directions and implications of this investment: nanoelectronics, optoelectronics and magnetics; advanced health care, therapeutics, and diagnostics; nanoscale processes for environmental improvement; efficient energy conversion and storage; microcraft space exploration and industrialization; bio-nanosensor devices for reduction of communicable diseases and biological threats; economical and safe transportation; and national security.

The important message for the Air Force is that extensive efforts are under way worldwide in micro- and nanotechnologies. It will therefore be essential to be selective in investing the relatively small Air Force research and development resources and to couple them effectively to the results of the many extramural efforts.

MAJOR AREAS OF OPPORTUNITY

The committee developed a taxonomy to codify the many micro- and nanotechnology opportunities and to assess their relative importance to the Air Force and the appropriate levels and kinds of efforts the Air Force should undertake. A very high level description of this taxonomy is presented in Table ES-1. A more detailed version that also serves as a guide to the body of the report is presented as Table ES-2.

ENABLING MANUFACTURING TECHNOLOGIES

The microelectronics, computer, and information revolutions can trace their success to several technological roots. The monolithic integration of transistors into functional blocks—which are further integrated to form microprocessors, memories, and other integrated circuits—is among the main reasons for the ever-increasing functionality. Mass production of integrated circuits by batch fabrication with high yields has led to both the declining costs per function and the

TABLE ES-1 Recommended Air Force Roles in Micro- and Nanotechnology Research

Topic	Air Force (AF) Takes the Lead	Air Force Participates
Information technology (computing, storage, communications, software)	Radiation-hard electronics, mission software concepts, AF-specific communications	Selected areas
Sensors and sensor systems	Electromagnetic; hyper- and multispectral	Multimodal, distributed, inertial
Biologically inspired materials and systems		Selected areas
Structural materials	AF-specific areas (e.g., low observability)	Micro- and nanostructures
Aerodynamics, propulsion, and power	Microsystems for aerodynamic sensing and control	Micropropulsion and propellants, energetics
Directed (lithographic) manufacturing		Selected areas
Self-assembly manufacturing		Selected areas
Component assembly		Selected areas
Packaging		For AF-specific needs
Reliability	AF-specific areas (e.g., space)	

NOTE: Where Air Force requirements far outpace academic/other government/commercial sector requirements, a lead role is recommended. Where there will be substantial academic and commercial sector developments, participation is recommended to leverage those developments and to apply the necessary Air Force-specific overlay. More detail on selected areas is provided in Table ES-2 and in the body of the report.

remarkable reliability of integrated circuits. A great triumph of microelectronics has been the high-yield manufacturing of reliable 100-million-part assemblies at an affordable cost. The possibility with nanotechnology is reliability of even more complex assemblages, and the commensurate challenge is the integration of an ever-widening array of materials and functionalities.

A very significant problem for the Air Force is the transitioning of scientific and technological developments across the very wide gap between the laboratory and the field. An example is the digital micromirror device that powers today's

TABLE ES-2 Taxonomy of Micro- and Nanotechnology Research Areas and Their Relevance to the Air Force

Development Area	Micro- and Nanotechnology
Information technology	
Electronics—hardware	Scaled CMOS
	Single-electron transistors
	Spin-based electronics
	Molecular electronics
	Carbon nanotube electronics
	Quantum interference devices
	Vacuum microelectronics
	Space electronics
Computing—architectures	Cellular automata
	Quantum computing
	Artificial brains with natural intelligence
Digital storage	Magnetic storage, hard disks
	Magnetic storage, MRAM
	Nanoindent storage
	Molecular memory
Communication	Secure communications (quantum encryption)
	Optical devices (e.g., semiconductor lasers)
	Electronic confinement—quantum wells, wires, and dots in III-V materials
	Optical confinement—fibers, waveguides, and photonic crystals
	MOEMS optical switches
	MEMS RF switches
	Bioinspired
Information processing	Data fusion
	Distributed and autonomous systems
	Software and codesign
Sensors	Distributed sensors and swarms emergent behavior
	Electromagnetic sensors, UV to RF
	Hyper- and multispectral sensing
	Navigational sensors (MEMS)
	Magnetic field sensors
	Chemical/biological threat detection
	System-status sensing
Bioinspired materials and systems	Sensors
	Materials
	Computing, communications, and information processing
	Enhanced human performance

EXECUTIVE SUMMARY

Air Force Investment Priority			Market Driver			Time Frame			Product Status		
L	M	H	C	Mix	Mil	N	M	L	R	D	M
X			X			X	X	X	X	X	X
X	X		X					X	X		
X	X		X					X	X		
X	X		X					X	X		
X	X		X					X	X		
X	X		X					X	X		
	X		X			X	X		X	X	X
		X		X		X	X	X	X	X	X
X			X					X	X		
X	X		X					X	X		
X	X		X					X	X		
X			X			X	X	X	X	X	X
	X			X		X	X		X	X	
X			X				X	X	X		
X	X		X					X	X		
	X		X			X	X		X		
	X		X			X	X	X	X	X	X
	X		X			X	X	X	X	X	X
	X			X		X	X	X	X	X	X
X			X			X	X		X	X	
	X	X	X			X	X		X	X	
X			X				X	X	X		
		X		X		X	X	X	X	X	
		X		X			X	X	X		
		X	X			X	X	X	X		
		X		X			X	X	X	X	
		X		X		X	X	X	X	X	X
		X		X		X	X	X	X	X	X
		X	X			X			X	X	X
		X		X		X			X	X	X
X			X			X	X	X	X	X	
	X		X			X	X	X	X	X	
X	X		X					X	X		
X	X		X				X		X	X	
X				X				X	X		
X			X			X			X		

continues

TABLE ES-2 Continued

Development Area	Micro- and Nanotechnology
Structural materials	
Lightweight/high-strength materials	Nanoscale grain size
	Composite/designer materials
Improved coatings	Friction and wear reduction
	Low maintenance
Multifunctional structures	Incorporating active elements—actuators
	Self-healing structures
	Low-observability structures
Materials for MEMS/NEMS	Low adhesion and reliability
Aerodynamics, propulsion, and power	
Flight vehicle aerodynamics	MEMS actuators on airfoil
Air-breathing propulsion and power	MEMS flow controls
	Distributed sensing and actuation for control of turbine instabilities
	MEMS-based propulsion
	MEMS-based power sources
Launch vehicle propulsion	Nanocoatings for fuel components
	Nanopowder aluminum propellants
	MEMS liquid rocket engines
Spacecraft propulsion	Micro thrusters
	Digital thrusters
	Field-ionization electric propulsion
	Tether electric propulsion
Space power	Semiconductor solar cells
	Nanostructured battery materials
Enabling manufacturing technologies	
Fabrication	EUV lithography
	Optical lithography
	Nanoimprint lithography
	Self-assembly
Assembly	Micro- and nano pick and place
	Bioinspired
Packaging	Micro-and nanosystems
	Embedded devices
Reliability	Mixed material systems
Commercialization	Foundries

NOTE: The recommended Air Force investment priorities are classified as low (L), medium (M), and high (H). The market driver refers to the principal impetus for product development—the commercial sector (C); mixed (mix), leveraging the commercial sector with military-specific requirements; or the military (mil). The time frame in which the technology is likely to be ready for Air

EXECUTIVE SUMMARY

Air Force Investment Priority			Market Driver			Time Frame			Product Status		
L	M	H	C	Mix	Mil	N	M	L	R	D	M
	X				X	X			X	X	X
	X				X		X	X	X	X	
		X			X	X	X	X	X	X	
	X				X	X	X	X	X	X	
X					X		X	X	X		
	X				X			X	X		
	X				X	X	X		X	X	
	X			X		X	X	X	X	X	X
	X				X			X	X		
	X				X		X	X	X		
X				X			X		X	X	
	X				X		X	X	X		
	X			X			X	X	X		
		X			X		X	X	X		
		X			X	X			X	X	
		X			X		X	X	X		
		X			X		X		X		
		X			X		X		X		
	X				X		X		X		
	X				X		X		X		
X				X		X	X	X	X	X	X
X	X			X			X		X	X	
X			X				X		X	X	
X				X		X	X		X	X	X
	X		X				X		X		
X	X			X			X	X	X		
X			X					X	X		
X			X					X	X		
	X		X			X	X	X	X	X	X
	X	X	X			X	X	X	X	X	
	X		X			X	X	X	X	X	X
	X			X		X	X		X	X	

Force applications is given as near term (N, 0–10 years), medium term (M, 10–20 years), and long term (L, 20–50 years). Product status refers to the stage of industrial development: research (R), development (D), or manufacturing (M). Technologies are presented in the order of their appearance in the body of the report.

ubiquitous digital projectors. The companies involved in the development of this technology had to invest many resources, and about 20 years, to take this device from the laboratory bench to the movie theatre. This serves to underscore the difficulty of the task—and the need to leverage all of the available resources, both within and outside of the Air Force, to make the transition.

AIR FORCE MICRO- AND NANOTECHNOLOGY PROGRAMS AND OPPORTUNITIES

The Air Force Research Laboratory (AFRL) is making a focused investment in micro- and nanotechnologies. There is a planning process under way, at the level of the AFRL Chief Technologist, to collect the existing programs of micro- and nanotechnology research within AFRL and turn them into a coordinated plan. However, this is a task made more difficult by declining budgets. For example, science and technology (S&T) (6.1-6.3) funding within the Air Force has decreased both in absolute terms and relative to the other Department of Defense (DoD) Services. Over the period 1989-2000, Air Force S&T funding decreased by almost 50 percent, from $2.7 billion in 1989 to $1.4 billion in 2000, in constant-year dollars. Over the same period, the Army and the Navy S&T budgets increased by 13 percent and 40 percent, respectively. Of significant concern is the 6.1 basic research budget, which focuses on long-range research. From 1989 to 2000, the Air Force investment in basic research declined by $39 million in real terms, a decrease of 15 percent.

The declining Air Force S&T budget poses several significant challenges: first, which of the many nanotechnologies to support with a limited budget, and how; second, how to leverage extramural nanotechnology developments; and third, as these micro- and nanotechnologies approach maturity, how to transition them into hardware and operational systems. These challenges are made more difficult by the increasing dominance of the commercial sector in product and manufacturing directions, sometimes with requirements very different from those of the military. It will be too expensive for the military to maintain a parallel manufacturing effort across the entire spectrum of micro- and nanotechnology, so it will have to adopt as much as possible from the commercial sector, applying its own overlay and integration and only selectively developing military-specific products.

Meeting these challenges will require a sustained effort by the Air Force. It will require leadership at the general officer level as well as at the highest levels of AFRL. It will require a cadre of AFRL staff who are full-fledged members of the broader micro- and nanoscience research community. It will require stronger ties to academia and other research efforts. It will require focused efforts, benchmarked against the best in the world, both in basic research and in multidisciplinary subsystems demonstrations that force participants to cross-fertilize each other's thinking and address real-world problems.

AIR FORCE PLATFORM OPPORTUNITIES IN MICRO- AND NANOTECHNOLOGIES

The committee found that advances in micro- and nanotechnologies will be relevant to all six of the core competencies within the Air Force strategic plan: aerospace superiority, information superiority, global attack, precision engagement, rapid global mobility, and agile combat support. Some potential systems consequences, organized around Air Force platforms, are highlighted in Table ES-3. This table ties together the core competencies, the systems applications, and the technology advances. Pursuit of any of these specific systems opportunities requires detailed considerations well beyond the scope of the present study.

FINDINGS AND RECOMMENDATIONS

The committee offers a number of critically important, broadly applicable findings and recommendations, which are presented below. Findings and recommendations related to technological developments in micro- and nanotechnologies are listed first (T), followed by policy recommendations (P). Findings and recommendations are presented in a logical flow. The numbering does not represent a rank ordering but simply serves to identify the findings and recommendations. In addition, a number of more specific findings and recommendations were developed that are listed in the body of the report and collectively in Chapter 7.

Technology Findings and Recommendations

Four overarching themes emerged from the committee's study of the implications of emerging micro- and nanotechnologies: increased information capabilities, miniaturization of systems, new materials resulting from new science at these scales, and increased functionality and autonomy (T8). The following findings and recommendations attempt to capture the essence of these themes with some specificity. The increased information capabilities flow from near-term continuation of the scaling of silicon electronics (T1) and from new and alternative concepts arising from nanotechnology research (T2). Biological science, both as inspiration (biomimetics) and as a functional contributor, offers new opportunities (T3) that build on and complement traditional sensing, computing, and communications approaches. Increased information capabilities and miniaturization together will make possible large distributed arrays of sensors (sensor swarms) on combinations of fixed and movable platforms. These array systems will exhibit new or emergent properties significantly different from those of individual components and will allow increasingly autonomous operation of Air Force systems (T4). Harnessing the capabilities of microelectromechanical systems to propulsion and aerodynamics will allow the miniaturization of air and space platforms (T5). Maximizing the utility of these increased capabilities in

TABLE ES-3 Selected Mission and Platform Opportunity Areas

System Type	Science and Technology Area					Selected Mission and Platform Opportunities	Time Scale[a]	Air Force Critical Future Capability					
	Information Technology	Sensors	Bioinspired Materials and Systems	Structural Materials	Aerodynamics, Propulsion, and Power			Aerospace Superiority	Information Superiority	Global Attack	Precision Engagement	Rapid Global Mobility	Agile Combat Support
Space vehicles and systems	X	X	X	X	X	*Distributed satellite.* Self-sustaining nano-satellite arrays/swarms to monitor, report status, and take action	M-L	X	X	X		X	X
	X	X	X		X	*Integrated spacecraft.* Highly integrated, reprogrammable, reconfigurable systems	M	X	X	X	X		X
				X	X	*Micro launch vehicles.* Low-cost, launch-on-demand tactical space systems	M-L	X	X	X	X	X	

EXECUTIVE SUMMARY

Category						Description	Timeframe[a]					
Weapon systems	X	X	X	X	X	Miniaturized ballistic missiles. Rapid global-reach system enabled by microtechnologies	M	X	X	X		X
	X	X	X	X	X	UAV-launched ABM boost-phase interceptors. Micro- and nano-enabled small missile interceptors	M	X		X	X	
	X	X	X	X		Air-to-air and air-to-ground weapons. Missiles and bombs with significantly reduced weight, size, and cost through miniaturization with better performance	M	X		X		X
Air vehicles and systems	X	X	X	X	X	Micro air vehicles. Low-cost, ubiquitous, autonomous surveillance and reconnaissance systems and microdecoys; cooperative behavior of swarms of vehicles	N-M-L	X	X	X		X
	X	X	X	X	X	MEMS-based active aerodynamic flight control. Microsensing and control of air flow combined with new materials for enhanced flight efficiency	M-L	X		X	X	X

NOTE: Also indicated is their relation to micro- and nanotechnology S&T areas, as discussed in Chapter 3, and to Air Force–defined critical future capabilities.

[a]N, near term; M, medium term; L, long term.

information technology, biomimetics, individual sensors and sensor swarms, and MEMS actuators for the Air Force will demand specific attention to system design, architecture, and software for system implementation (T6). Because of the wide range of new capabilities being enabled, the trend toward merging heterogeneous materials systems and toward expanding the range of materials in micro- and nanoscale devices and systems is inexorable (T7).

Finding T1. Further miniaturization of digital electronics with increased density (~128×) is projected by the integrated circuit industry over the next 15 years based on continued scaling of current technology.

Recommendation T1. The Air Force should position itself to take advantage of the advances predicted by the Information Technology Roadmap for Semiconductors.

Finding T2. In anticipation of an ultimate end to the historical scaling of today's integrated circuit technology, many new and alternative concepts involving nanometer-dimensioned structures are being examined.

Recommendation T2. Exploration of the scientific frontiers involving new procedures for fabrication at nanodimensions and new nanoscale materials, properties, and phenomena should be supported. The Air Force should track, assimilate, and exploit the basic ideas emerging from the research community and continue to support both intra- and extramural activities.

Finding T3. Biological science offers new opportunities in nanotechnology systems, especially for sensors, materials, communications, computing, intelligent systems, human performance, and self-reliance.

Recommendation T3. The Air Force should closely monitor the biological sciences for new discoveries and selectively invest in those that show a potential for making revolutionary advances or realizing new capabilities in Air Force-specific areas.

Finding T4. Large, distributed fixed arrays and moving swarms of multispectral, multifunctional sensors will be made possible by emerging micro- and nanotechnology, and these will lead to significant fundamental changes in sensing architectures.

Recommendation T4. The Air Force should develop balanced research strategies for not only the hardware but also the requisite software and software architectures for fixed arrays and moving swarms of multispectral, multifunctional sensors.

Finding T5. Emerging microtechnology offers new opportunities in propulsion and aerodynamic control, in particular in (1) distributed sensors and actuators on both macro-aerodynamic surfaces and macro-aeropropulsion units and (2) new, scalable, miniaturized and distributed aero- and space-propulsion systems.

Recommendation T5. The Air Force should move decisively to develop new research and development programs to bring microtechnology to both macro- and microscale propulsion and aerodynamic control systems.

Finding T6. The Air Force strategic nanotechnology R&D plan, as presented to the committee, is focused on hardware concepts without appropriate consideration of total systems solutions.

Recommendation T6. The Air Force should take seriously the importance of co-system design as a critical implication of continued miniaturization and should invest in the algorithm, architecture, and software R&D that will enable the codesign of hardware and software systems. This should be undertaken along with a projection of the advances that will be made in hardware.

Finding T7. Integration of micro- and nanoscale processes and of different material systems will be broadly important for materials, devices, and packaging. Self-assembly and directed assembly of dissimilar elements will be necessary to maximize the functionality of many micro- and nanoscale structures, devices, and systems. Achievement of high yields and long-term reliability, comparable to those of the current integrated circuit industry, will be a major challenge.

Recommendation T7. The Air Force should monitor progress in self- and directed-assembly research and selectively invest its R&D resources. It will be critical for the Air Force to participate in developing manufacturing processes that result in reliable systems in technology areas where the military is the dominant customer—for example, in sensors and propulsion systems.

Finding T8. Four overarching themes emerge from the advance of micro- and nanotechnologies—increased information capabilities, miniaturization, new engineered materials, and increased functionality/autonomy. These themes could have a significant military impact by enabling new systems approaches to Air Force missions.

Recommendation T8. The Air Force should continue to examine new systems opportunities that may emerge from the successful development of micro- and nanotechnologies and use these studies to help focus its applied research and development investments in these technologies.

Policy Findings and Recommendations

The Air Force critically depends on advanced technology to accomplish its missions. In order to maintain the U.S. competitive technology advantage over the long term, the Air Force must maintain a stable, robust, and effective research, development, testing, and evaluation (RDT&E) program. The Air Force is currently underinvesting in this critical area and has not maintained the stability necessary for sustained progress, thereby shortchanging its future and that of the nation (P1). An important new development is that the commercial sector now overshadows the military market. This means that product development is driven by commercial, not military, requirements. The DoD cannot, however, rely solely on commercial R&D and products to satisfy its needs (P2). Micro- and nanotechnologies are going to play a major role in future Air Force systems, as detailed in the technical sections of this report (the basis for findings and recommendations T1-T8 above). The Air Force Research Laboratory has initiated a planning process to enhance its effectiveness in this all-important area (P3), but more needs to be done to strengthen the Air Force's internal programs and to ensure that they assimilate and leverage the results of the very extensive programs under way throughout the worldwide scientific community (P4).

Finding P1. Both overall DoD and—even more—Air Force policies have de-emphasized R&D spending to the detriment of DoD and Air Force long-term needs. The Air Force relies heavily on the technological sophistication of its platforms, systems, and weapons. Its ability to meet its long-term objectives is critically dependent on a strong and continuing commitment to R&D.

Recommendation P1. The Air Force must significantly increase its R&D funding levels if it is to have a meaningful role in the development of micro- and nanotechnology and if it is to be effective in harnessing these technologies for future Air Force systems.

Finding P2. The military market for many micro- and nanotechnologies (e.g., advanced computing, communications, and sensing) is small in comparison with commercial markets. Yet, the Air Force and DoD have mission-specific requirements not satisfied by the commercial market. Military-specific applications will not be supported by industry without government and Air Force investment, particularly for basic research.

Recommendation P2. The Air Force should concentrate its efforts in micro- and nanotechnology on basic research at the front end and on Air Force-specific applications at the back end.

Finding P3. The Air Force has recognized the importance of micro- and nanotechnologies for its future capabilities and has begun a planning process to maximize the benefits of its in-house and extramural research programs. Strong leadership will be necessary to ensure maximum benefit from the Air Force Research Laboratory research programs.

Recommendation P3. It will be critical to continue the planning for micro- and nanotechnologies at the highest levels of the Air Force Research Laboratory (AFRL). AFRL should also strengthen its external review processes to assist the leadership and to ensure that its work is well coordinated with national efforts. The Air Force should coordinate its initiatives with other federal agencies and work to build collaborative programs where appropriate.

Finding P4. The committee perceived a lack of consistency in the quality of current in-house Air Force programs and in the benchmarking of those programs against the large number of programs under way throughout the world.

Recommendation P4. Considering that micro- and nanotechnology is a new and rapidly emerging interdisciplinary field, the Air Force should critically evaluate its efforts in micro- and nanotechnology to select areas of strong potential payoff for Air Force missions and to sustain the highest-quality program.

REFERENCES

1. International Technology Roadmap for Semiconductors. 2001. Available online at <http://public.itrs.net/> [July 1, 2002].
2. MEMS Industry Group. 2001. 2001 Annual Report. Available online at <http://www.memsindustrygroup.org/arord01.htm> [July 2, 2002].
3. National Science and Technology Council. 2000. National Technology Initiative: The Initiative and Its Implementation Plan. Available online at <http://www.nano.gov/nni2.pdf> [July 2, 2002].

1

Introduction

BACKGROUND

The prediction of the progress of technology over long periods is an uncertain exercise. The temptations are to be either too conservative, acknowledging the current limitations of technology and not foreseeing the breakthroughs in conception and capability that will inevitably occur, or too exuberant, brushing aside real physical limitations in an excess of futuristic zeal. Such a challenge particularly applies to nanotechnology, which is an exciting and relatively unexplored scientific and technological frontier offering many new insights and applications but at the same time evoking much speculation and hyperbole. To cite Shermer in one of the many recent journal overviews of nanotechnology, "The rub in exploring the borderlands is finding that balance between being open-minded enough to accept radical new ideas but not so open-minded that your brains fall out."[1] From an applications perspective, the combination of microtechnologies and nanotechnologies offers a particularly powerful combination for future Air Force missions and deserves careful consideration.

Two particularly pervasive themes of microtechnology, now extending into nanotechnology, have been miniaturization of electronic systems and the resulting increase in information density. The miniaturization trends of the last 50 years will undoubtedly continue and even accelerate over the next 50 years. In 1950, we had five-transistor radios, and computers were vacuum-tube-filled rooms with a very limited mean time between failure, available only to governments and large corporations. Today we have inexpensive, 100-million-transistor computer chips in our homes, which we replace not because they have failed, but

INTRODUCTION   21

because the technology has advanced. We complain about the demands of ubiquitous connectivity as we attach cell phones to our belts and of information overload as we put more and more material on our Internet servers. Over this time span, transistors have gone from macroscopic, ~1-millimeter junction-length devices, to ~90-nanometer gates in the latest commercial chips and to ~10-nanometer gates in laboratory devices. This linear scaling clearly must end as devices approach the size of atoms (~0.2 nanometers). This does not, however, mean that progress in electronics and information technology will come to a halt. The integrated circuit paradigm that has enabled this dramatic scaling improvement is a planar, two-dimensional concept based on an interconnection of three-terminal switching elements (transistors).[2] Moving to a volumetric approach, new materials, and different computing strategies will probably allow continuation and even acceleration of the capabilities and function per weight/volume/power of electronics. The practical success of miniaturization has been the result of the accompanying dramatic reduction in cost per function achieved by the integration of so many electronic devices onto single chips and using parallel, or batch, fabrication technologies to allow this cost scaling.

Less well understood is the acceleration in other micro- and nanotechnologies, which is being driven by miniaturization and is contributing to the increasing density of information transmitted, stored, and processed. The growth in magnetic information storage in recent years has been even more rapid than growth in electronic information processing.[3] Advances in magnetic memory storage range from new giant magnetoresistive nanoscale layered materials to read heads flying 10 nanometers over the surface of magnetic discs moving at speeds of 20 meters/second. To appreciate the challenge in control of tolerances for this technology, scaling to the macro world by the relative lengths of a magnetic read head and an F-18 jet fighter would correspond to flying the F-18 only 100 micrometers above the ground, which has been polished to a smoothness of 10 micrometers and staying on course within an accuracy of 100 micrometers. Optical information transmission has also been increasing at growth rates comparable to that for magnetic memories, aided by control of materials—for example, in optical fibers with ultrahigh-purity microscale cores and semiconducting lasers with nanoscale quantum wells.

Mechanical devices at the microscale and below promise to further extend the reach of miniaturized technologies. Microelectromechanical systems (MEMS) build on the manufacturing paradigm of microelectronics and offer the promise of large-scale batch fabrication at low cost. Currently this emerging technology is primarily focused on simple devices such as inertial sensors for air bag release in automobiles and microscale mirrors for optical projection and switching. However, future applications of MEMS for airfoil control, inertial sensing, or satellite maneuverability could significantly broaden the scope of this technology. The integration of MEMS technologies with electronics and optics is also being explored for chemical sensing, so-called lab-on-a-chip systems. Indeed, the current status of this technology appears to be similar to that of microelectronics some 25 years ago.

The emerging breadth of microscale technologies (mechanical, optical, magnetic, chemical, and biological, as well as electronic) and the promise of future nanoscale technologies suggest that revolutionary advances in systems are likely. Miniaturization and high information density will be particularly important where performance requirements place weight and size at a premium. The potential of low cost, if achieved, implies the ubiquitous use of devices, as is now happening in microelectronics with the embedding of computer chips throughout systems. The widespread ability to embed high information density in combination with local detection, processing, and response in small packages will allow large networks of distributed systems and increasingly autonomous systems. The overarching theme that emerges is increased functionality and autonomy of systems. Low cost, ubiquitous, distributed systems will raise new questions such as the role of autonomous control and decision making and the integration of such system capabilities into military conduct of operations.

STATEMENT OF TASK

This study was requested by the Deputy Assistant Secretary of the Air Force for Science, Technology and Engineering. The Committee on Implications of Emerging Micro- and Nanotechnologies, established by the National Research Council, was asked to perform the following tasks:

- Conduct a study to examine the role that emerging micro- and nanotechnologies can play in improving current Air Force capabilities and enabling new weapons, systems, and capabilities.
- Assess the current state of the art in micro- and nanotechnologies.
- Discuss how current and future Air Force mission capabilities may be impacted or enabled by these technologies.
- Review the current Air Force and Department of Defense (DoD) investment strategies and the Air Force plan of execution in micro- and nanotechnologies for adequacy; recommend directions for accelerating the operational success of these technologies in Air Force missions.
- Recommend research initiatives that are needed to explore promising micro- and nanotechnologies.

WHAT WE MEAN BY "MICRO" AND "NANO"

In undertaking this study, the committee decided not to put hard size limitations on micro- and nano- objects and technologies. It understands these concepts as relating roughly to scale but also as having significant differences in underlying physical and chemical mechanisms. There is no hard line between micro and nano, but there are some clear differences in the way the scientific and technology communities approach these regimes. It is very difficult to come up with sufficiently inclusive definitions for these concepts that hold across the multiple disciplines that are being explored, but it is not as difficult to decide in an individual case which category it fits into—in some sense, "you know it when you see it" holds here as well as in less noble realms of human activity. So, the committee defines micro and nano by example (see Figures 1-1, 1-2, and Box 1-1).

Science and technology are always heavily intertwined and impossible to discuss, or indeed to advance, independently. Understanding the science enables the technology, and harnessing the technology allows further advances in the science. For conciseness in this report the committee speaks of micro- and nanotechnology, but this should always be understood to mean both micro- and nanoscience and micro- and nanotechnology.

Microtechnology is characterized by a top-down fabrication paradigm, where the starting point is macroscopic and material is added or taken away in processes such as lithography to define patterns on surfaces, etching to remove material, and deposition to add material and thus allow complex structures to be made. The integrated circuit is an example of this paradigm. The starting point is an almost perfect wafer of silicon. Areas are defined on this wafer for introducing electrically active dopants, for adding various electrodes (source, gate, and drain contacts of transistors), and for making interconnections. When it was first conceived

FIGURE 1-1 Model of a MEMS safety switch. SOURCE: Sandia National Laboratories. 2002. Available online at <http://mems.sandia.gov/scripts/images.asp> [June 21, 2002].

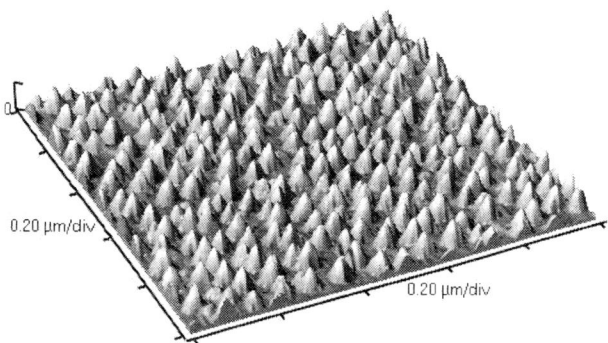

FIGURE 1-2 Atomic force microscopic image of InAs quantum dots. SOURCE: Courtesy of Kevin Malloy, University of New Mexico.

in the late 1950s by Jack Kilby at Texas Instruments and Bob Noyce at Intel, the integrated circuit was a response to the difficulty of reliably packaging together numbers of transistors, resistors, capacitors, and other circuit elements to make large-scale circuits. The technology was known initially as "the monolithic solution."[4] MEMS devices, which now cover a very broad range of application, from accelerometers and angular rate sensors to switches to infrared bolometer focal plane arrays, are further examples of what we are calling micro. Typical scales range from a few hundreds of micrometers down to one micrometer and less. At the microscale, objects have greatly reduced inertia, and turbulence, convection, and momentum become negligible. At this scale, the surface and interface properties of materials begin to play an increasingly dominant role in the behavior of structures.

A defining feature of the nanoscale is that there is a qualitative difference in material behavior, which does not scale from the macro and micro scales. New physics and chemistry come into play. Another way to say this is that dimensions, in addition to composition and structure, impact material properties in nanoscale materials. At least two factors dominate this transition. The first is that dimensions in the nanometer regime approach characteristic (quantum) wavefunction scales of excitations in the material—electrons and holes, photons, spin waves, and magnons, among others. The second factor is the very large surface to volume ratio of these structures, which means that no atom is very far from an interface; atomic forces and chemical bonds dominate.

The first factor is the domain of quantum physics. Electronic wave functions (the de Broglie length) in semiconductors are typically on the order of 10 to 100 nanometers. The solid-state physics community has long been exploring the properties of quantum wells, in which one dimension (the growth direction) is on this scale. Modern telecommunications is based on semiconductor quantum-well lasers that exploit the unique properties of these structures. More recently, attention

BOX 1-1
A Matter of Scale

We are used to thinking in a linear world. Changes in scale by many factors of 10 challenge both our intuition and our imagination. In going to the world of micro- and nanoscale phenomena we move to smaller dimensions by factors of 1,000, the micrometer being one one-thousandth of a millimeter (the diameter of the tip of your pen) and the nanometer being a million times smaller than the pen tip, or one one-thousandth of a micrometer. Figure 1-1-1 shows examples of the impact of the nanoscale on macroscopic objects for both biological and artificial systems. To visualize this scale change for everyday objects, consider your morning cup of coffee. The width of the mug is about 80 millimeters (mm). Now consider the diameter of a human hair, which is typically 50 micrometers (mm), a reduction in scale by about 1,000. To reach the nanoworld yet another reduction of 1,000 is needed. An object 50 nanometers (nm) high corresponds to a stack of about 200 atoms. If packed without space between them, a billion of these nano-objects would fit within a 50- × 50- × 50-micrometer cube!

The speed of moving objects is another way to visualize large changes in scale. Consider a baby crawling. A person walking goes about 10 times as fast, a car traveling at 60 mph is 100 times faster, and a jet fighter at the speed of sound is 1,000 times faster than the crawling baby. Factors of 1,000 in moving between the macro-, micro-, and nanoworlds are truly large changes that challenge our intuitive capabilities.

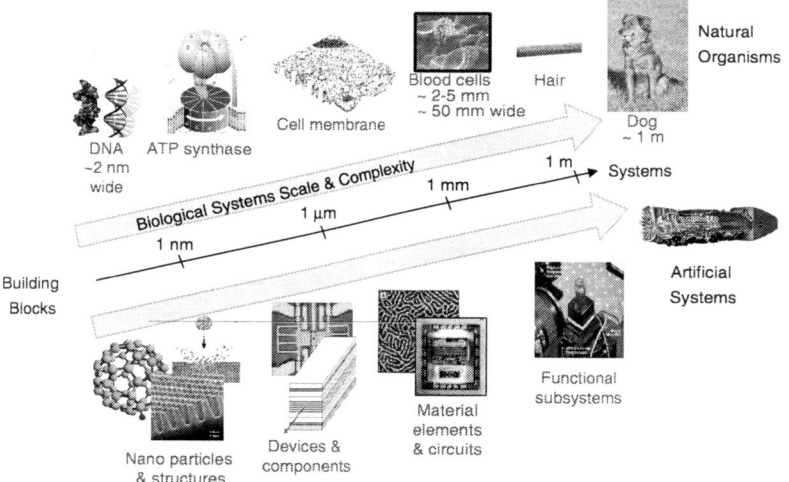

FIGURE 1-1-1 Dimensional scale. SOURCE: Wilson, B. 2001. AFRL Nano Science and Technology Initiative. Briefing by Barbara Wilson, Chief Technologist, Air Force Research Laboratory, to the Committee on Implications of Emerging Micro- and Nanotechnologies, National Academy of Sciences, Irvine, Calif., December 18.

has turned to quantum dot structures that have all three dimensions in this regime. In some sense, these are "designer" atoms and molecules that can be engineered to provide the needed functionality. Another example is the wavelength of visible light, which is 400 to 800 nanometers. When periodic structures are created in optical materials at these dimensional scales by varying the dielectric constant, the propagation of light can be strongly influenced in analogy to electrons in semiconductors. While these properties are only now being explored, the possibilities include confining and steering light down to unprecedented small scales and creating low loss-optical devices such as near-thresholdless lasers.

The second factor is a consequence of the large surface areas and unique chemical reactivity of nanostructures. This is the basis for much of the excitement at the juncture of nano- and biotechnologies. The information stored in the genome and the exquisite selectivity of biochemical interactions based on chemical recognition and matching are examples of nanoscale properties where the interfaces play a determining role. Nanoparticles have size-dependent chemical and electronic structure, reactivity, etc. that can be exploited to produce improved catalysts as well as electronic, magnetic, optical, and biomaterials.

Materials constituted of nano particles are different from bulk materials and different from molecules. An easy characterization is to say that nanoscale objects contain a large (more than a simple molecule) but countable (for example a box of 100 atoms on a side containing 1 million atoms) number of atoms. With our increasing ability to fabricate structures with well-defined nanoscale features, new materials are emerging that promise both evolutionary and unexpected new properties. Another major thrust of nanoscale research is integration, where the aim is to preserve the unique properties of nanoscale structures as they are incorporated into macroscopic objects.

Nanotechnology is generally anticipated to require a fundamentally different approach to fabrication than microtechnology. Whereas microscale structures are typically formed by top-down techniques such as patterning, deposition, and etching, the practical formation of structures at nanoscale dimensions will require an additional component—bottom-up self-assembly. This is the process whereby structures are built up from atomic- or molecular-scale units into larger and increasingly complex structures—as is widely used by biological systems. In practice some combination of top-down (lithographic) and bottom-up (self-assembly) techniques likely will be necessary for the efficient manufacturing and integration of nanoscale systems. Many tools now exist for investigating structure and properties at the nanoscale, including scanning tunneling probes, electron microscopies, and various diffraction techniques. An important development in nanoscale tools occurred in 1981 with the introduction of the scanning tunneling microscope for imaging individual atoms on surfaces. This development, which earned Bennig and Rohrer the Nobel Prize in Physics, allowed the imaging and manipulation of single atoms and set the stage for an entire family of scanning microscopy, with atomic force microscopy being the most widely used.

As will be seen in the following pages, four overarching themes emerged from the committee's study of micro- and nanotechnologies:

- Increased information capabilities,
- Miniaturization of systems,
- New materials resulting from new science at these scales, and
- Increased functionality and autonomy.

These themes emerge as a natural consequence of the advances in micro- and nanotechnologies resulting from scaling to small size. They will have far-reaching consequences for Air Force missions.

Finally, the committee notes that not all things "nano" adhere to the usual nanometer dimensional scale, nanosatellites being a notable example. In this case nanosatellites have overall dimensions of many centimeters—the name evolved as a way to designate systems that are significantly smaller in a revolutionary way from today's large, expensive satellite technology (see Box 1-2). However, even here the basis for developing nano satellites is provided by advances in micro- and nanotechnologies.

REPORT ORGANIZATION AND METHODOLOGY

This report documents the committee's analysis, findings, and recommendations. The Air Force asked the committee to address short-term impacts as well as longer-term impacts, 20 to 50 years out. Both micro- and nanotechnologies were included because in combination they cover the near- and long-term trends in modern technology that will impact Air Force missions. These trends are most apparent in microelectronics and include the miniaturization of components, increased capability (information density), reduced cost per function, and increased reliability and ruggedness. Advances in microtechnology are evolving smoothly into other areas, such as MEMS for micromechanical components, and control at the nanoscale is helping to improve the performance of microscale systems. At the same time new, more revolutionary advances in materials, properties, and, ultimately, systems are emerging at the nanoscale.

In Chapter 2, "Expectations for Future Micro- and Nanotechnologies," a brief overview of current perspectives in micro- and nanotechnologies is presented. Chapter 3, "Major Areas of Opportunity," addresses advances in micro- and nanotechnology areas most relevant to the Air Force. The committee included sections on information technology, sensors, biologically inspired materials and systems, structural materials, aerodynamics, and propulsion and power. These are all areas that could be of great interest to the Air Force; however, they do not necessarily merit the same level of emphasis. In Chapter 4, "Enabling Manufacturing Technologies," the challenges and trends faced by the practical realization of micro- and nanoscale materials, components, and systems are dis-

BOX 1-2
Small Satellites: How Small Can We Go?

The term "nano" has taken on a different meaning in the context of satellites; yet these future miniaturized systems are firmly based on advances in micro- and nanotechnologies. Popular early in the space age because of payload limitations on launch vehicles, "microsatellites" are satellites with a mass between 10 and 100 kilograms. More recently, satellites with mass between 1 and 10 kilograms have been called "nanosatellites," while those with mass between 0.1 and 1 kilogram are now called "picosatellites." Even smaller are "femtosatellites," between 10 and 100 grams.

The first artificial Earth satellite, Sputnik-1, launched October 4, 1957, had a mass of only 83.6 kilograms. The continually improving payload capabilities of launch vehicles have led to ever-larger active satellites with greater spacecraft power and communications data rate.

For the most part, communications satellites have migrated from low Earth orbit (LEO—below 1,500-kilometer altitude) to fixed geosynchronous orbit (GEO—35,786-kilometer altitude), requiring relatively large and costly spacecraft. During the 1980s, integrated circuit and radio frequency communications technology advanced to the point where microsatellites in LEO could provide competitive communications and data relay support, including the support of military forces in the field. In 1990 the Defense Advanced Research Projects Agency launched two experimental 66-kilogram-mass multiple access communications satellites (MACSATs), providing store-and-forward communications at up to 2.4 kilobits per second.[1] The MACSATs were used for logistics communications in support of a Marine air wing in Desert Shield and Desert Storm. Today, technology has evolved to the point that nanosatellites and even picosatellites can perform complex scientific, communications, Earth observation, and satellite assistance missions. Figure 1-2-1 shows a modern nanosatellite fabricated by Surrey Satellite Technologies in the United Kingdom and designed to perform a satellite inspection mission.

---

[1] Martin, D.H. 1996. Communication Satellites, 1958-1995. El Segundo, Calif.: The Aerospace Press.

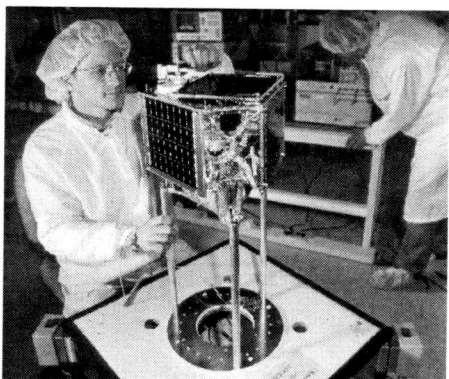

FIGURE 1-2-1 The SNAP-1 nanosatellite. Courtesy of Surrey Satellite Technology Limited, Centre for Satellite Engineering Research, University of Surrey, Guildford, Surrey, United Kingdom.

cussed. Chapter 5, "Air Force Micro- and Nanotechnology Programs and Opportunities," briefly summarizes the current investments by the Air Force in micro- and nanotechnologies and considers the role of Air Force science and technology in this area relative to the commercial sector. Chapter 6, "Opportunities in Micro- and Nanotechnologies," focuses on the systems implications of micro- and nanotechnologies and suggests areas for further consideration. Such mission considerations provide a methodology to focus on and prioritize investments for those technologies discussed in Chapters 3 and 4 that are most critical to the Air Force. Finally, Chapter 7, "Findings and Recommendations," provides a summary of the findings and recommends ways in which the Air Force might focus its attention and resources in the areas of micro- and nanotechnology.

REFERENCES

1. Shermer, M. 2001. Nano nonsense and cryonics. Scientific American 285(3): 29.
2. Reid, T.R. 2001. The Chip: How Two Americans Invented the Microchip and Launched a Revolution. New York, N.Y.: Random House, Inc.
3. National Research Council. 2001. Physics in a New Era: An Overview. Washington, D.C.: National Academy Press.
4. Reid, T.R. 2001. The Chip: How Two Americans Invented the Microchip and Launched a Revolution. New York, N.Y.: Random House, Inc.

2

Expectations for Future Micro- and Nanotechnologies

OVERVIEW OF CURRENT STUDIES

Visions for the future of nanotechnology and for its impact on technology in general are many. In this chapter, the committee discusses three: the International Technology Roadmap for Semiconductors (ITRS), the MEMS Industry Group 2001 Annual Report, and the National Nanotechnology Initiative (NNI).

International Technology Roadmap for Semiconductors

In 1992, the Semiconductor Industry Association (SIA) published a roadmap that charted the future of the industry by providing consensus predictions of future requirements and suggested solutions to anticipated problems with a 15-year time horizon. The industry roadmapping has evolved into a continuous process, with major reissues every 2 years. The most recent full update, the 2001 ITRS[1] includes input from semiconductor manufacturers worldwide and covers a 16-year period, divided into the near term (2001-2007) and the long term (2008-2016). Past SIA roadmaps have proven to be reliable predictors of future integrated circuit performance.

The growth in integrated circuit (IC) performance over the past 30 years has been both phenomenal and consistent. One performance measure, the number of transistors per chip, or per integrated circuit (see Figure 2-1), demonstrates the exponential growth predicted by Moore's law. This exponential growth has been the result of continually shrinking the dimensions of integrated circuit elements from tens of micrometers to tens of nanometers. Figure 2-2 shows how this

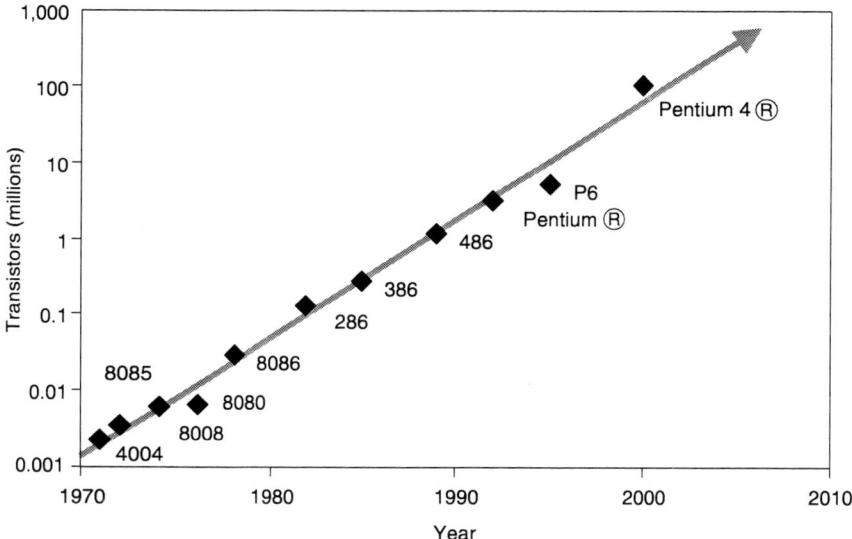

FIGURE 2-1 Integrated circuit growth. SOURCE: Adapted from Intel. 2002. Moore's Law. Available online at <http://www.intel.com/research/silicon/mooreslaw.htm> [May 31, 2002].

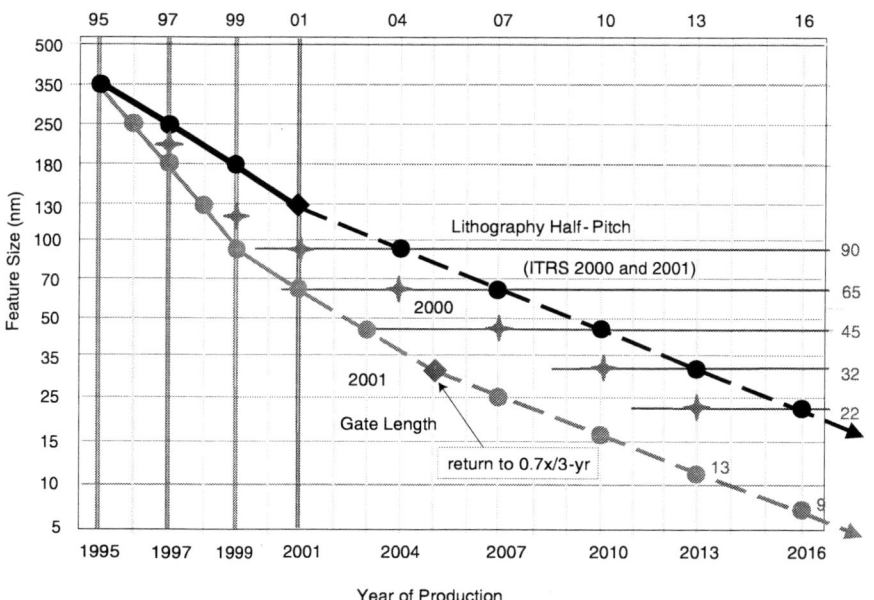

FIGURE 2-2 Lithography half-pitch feature size versus time. SOURCE: Adapted from the International Technology Roadmap for Semiconductors, 2001 edition. Austin, Tex.: International SEMATECH.

affects the minimum feature size (transistor gate length) required. Scaling has been employed to increase nearly all aspects of performance, including speed, power consumption per operation, and reliability. At the same time, the cost of the highest performance chips has remained approximately constant, and the cost per function has gone down dramatically, by over nine orders of magnitude.

The 2001 ITRS consists of target values for representative characteristics of circuits that are predicted to be in production each year from now until 2016. Examples include dynamic random-access memory (DRAM) half-pitch, multi-processor unit gate length, and junction depth. The manufacturability of each characteristic quantity is evaluated and assigned to one of three categories: manufacturable, currently being optimized; manufacturable, solutions are known; and, manufacturing solutions are not known. A point in the future where several quantities in the roadmap fall into the "manufacturing solutions are not known" category signals trouble and is called a red-brick wall. For example, a brick wall currently exists in 2007, at the end of the near-term period, indicating that research breakthroughs will be required to maintain the predicted schedule. Both new materials and advanced metal oxide semiconductor field effect transistor (MOSFET) structures may be required to break through this red brick wall. The longer-term challenge is to develop a manufacturable non-classical MOSFET structure to extend the technology to and beyond the end of the roadmap.

Table 2-1 shows the predictions of the 2001 ITRS for a few selected parameters that are characteristic of DRAM and microprocessor technologies. If the roadmap is achieved, over the period of 2001 to 2016, we will see a 128-fold increase in memory at three-fourths the cost and a 32-fold increase in microprocessor functionality and a 17-fold increase in microprocessor speed at one-fifth the cost. These huge gains in storage, functionality, and speed, as we shall see later in this report, have important implications for the Air Force.

No clear path is seen for extending the technology beyond the roadmap horizon of 2016. Indeed, a transition to an alternative technology may be needed before 2016. The 2001 ITRS notes as follows:[2]

> The horizon of the Roadmap challenges the most optimistic projections for continued scaling of CMOS (for example, MOSFET channel lengths of roughly 9 nm) . . . [and] it is difficult for] most people in the semiconductor industry to imagine how we could continue to afford the historic trends in process equipment and factory costs for another 15 years!

A new technology will be needed, because, beyond the numerous technological and economic difficulties arising with sub-10-nanometer minimum dimensions, MOSFET integrated circuit (IC) technologies cannot shrink indefinitely. Ultimately, the discrete atomic nature of matter limits the shrinkage. Before that, however, the quantum phenomena characteristic of nanotechnology become important, and scaling laws that have not taken such phenomena into account will prove problematical. For example, quantum mechanical tunneling of charge

TABLE 2-1 Predictions of 2001 ITRS for Selected Parameters

Characteristic	Year of Production					
	2001	2004	2007	2010	2013	2016
Memory						
DRAM feature size, ½ pitch (nm)	130	90	65	45	32	22
Generation at production	512M	1G	4G	8G	32G	64G
Cost/bit at production (packaged microcents)	7.7	2.7	0.96	0.34	0.12	0.042
Logic						
Minimum feature size (physical gate length) (nm)	65	37	25	18	13	9
Functions/chip (millions of transistors)	276	553	1,106	2,212	4,424	8,848
Cost/transistor, high performance, at production (microcents)	97	34	12	4.3	1.5	0.54
Local clock (MHz)	1,684	3,990	6,739	11,511	19,348	28,751
Power dissipation, high performance (W)	130	160	190	218	251	288

SOURCE: Adapted from the International Technology Roadmap for Semiconductors, 2001 edition. Austin, Tex.: International SEMATECH.

carriers becomes significant at 10 nanometers and dominates below 3 nanometers. In addition, at a few nanometers, about 50 percent of all atoms in a particle are surface atoms, so that electrical properties are no longer determined by solid-state bulk phenomena.

On the other hand, the same nanoscale phenomena offer the opportunity to develop whole new classes of devices based on the principles of quantum physics, which may propel integrated circuits toward and beyond the end of the current roadmap. FlashRAM, now widely used to store data in digital cameras and MP3 music players, uses quantum-mechanical tunneling of electrons through ~10-nanometer-thick dielectric layers to charge a memory storage element. The ITRS (2001) and the European Commission Technology Roadmap for Nanoelectronics (2000)[3] discuss a number of candidates for replacing classical CMOS. However, it may be that systems based on these new principles function in a manner significantly different from current computing systems and are more suited to alternative architectures. There is no established roadmap to the development of these technologies, even though one may speculate on their functionality. A possible path is illustrated in Figure 2-3.

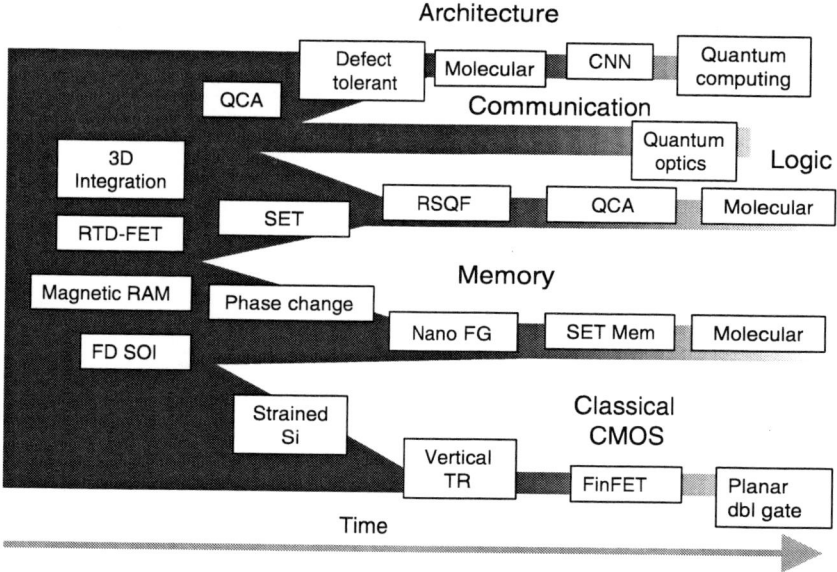

FIGURE 2-3 Possible roadmap. QCA, quantum cellular automata; CNN, cellular nonlinear networks; RTD–FET, resonant tunneling diode–field effect transistor; SET, single electron transistor; RSQF, rapid single quantum flux logic; nano FG, nano floating gate; SET mem, single electron transistor memory; FD SOI, fully depleted silicon on insulator; SI, silicon; vertical TR, vertical transistor; FinFET, a form of double gate transistor; planar dbl gate, planar double gate transistor. SOURCE: Adapted from the International Technology Roadmap for Semiconductors, 2001 edition. Austin, Tex.: International SEMATECH.

MEMS Industry Group 2001 Annual Report

The MEMS Industry Group is a nonprofit trade association of 22 member companies, established, in part, to "enable the exchange of non-proprietary information among members" and "provide reliable industry data that furthers the development of technology." Its annual report for 2001 discusses the key drivers and challenges expected by the industry over the next 20 years.[4] The committee briefly summarizes the findings here.

Starting with materials and fabrication approaches developed for the microelectronics industry, multiple miniaturized electronic and mechanical subsystems are integrated into a single microsystem. MEMS devices find applications in diverse industries because of their ability to perform complex functions and their relatively low cost, small size, low weight, high reliability, and good performance. The MEMS industry was estimated in 2000 to be between $2 billion and $4 billion in size (contrasted with about $180 billion for the IC industry) with 1.6 MEMS devices per person in use in the United States in 2001 and a compound annual rate of growth of 45 percent. MEMS production in the United States is currently concentrated on sensors and telecommunications, but there is a strong trend toward diversification.

The MEMS Industry Group's 2001 Annual Report foresees tremendous growth for the industry over the next 20 years. It discusses the expected roadmap for the industry in terms of the critical drivers and challenges expected, with driver defined as a force that "causes significant reactions in MEMS technology or commercialization." The advent of wireless networks and communications are a major driver for the introduction of MEMS into information technology. Drivers for the application of MEMS in health care are our aging population and our strong concern for a high quality of life.

Technological drivers include the introduction of new materials, e.g., plastics and magnetic and biocompatible materials, into the MEMS repertoire. Biotechnological drivers will include cell handling and living organisms. Other technological drivers involve interfaces between systems at different length scales, the need for low power operation, and the dramatic increase in computing power. Higher-fidelity modeling is necessary to ensure first-pass success. On the fabrication side, technological drivers include the use of magnetic and plastic materials within the process. A technological driver anticipated in the long range is the creation of a single fabrication process usable in a variety of electronic applications. The use of MEMS in extreme environments and for biomedical applications will be technological drivers for MEMS packaging. All of these applications would benefit enormously from advances in MEMS-specific design, modeling, and simulation tools.

The report also discusses technology challenges—that is, the known barriers to future progress in MEMS development and applications. First, MEMS fabrication equipment manufacturers are seen as being too few in number and too limited in diversity and quality. Similar problems are seen in packaging equipment for MEMS. Second, there is a need for standard (as opposed to custom) foundry services to facilitate testing and characterization by MEMS producers. Third, a library of reusable circuit designs needs to be developed to provide interfaces with MEMS. Fourth, packaging methods specific to MEMS need to be developed, and their effect on performance needs to be quantified. Other near- and intermediate-term issues concern communication and the need for a suitably trained workforce, e.g., MEMS-trained circuit designers. Over the longer term, there will be technological challenges relating to the stacking, assembly, lithography, and processing of three-dimensional geometries and the greater integration of electronics with MEMS.

The National Nanotechnology Initiative

In September 1999, the President's Interagency Working Group on Nanoscience, Engineering and Technology (IWGN) issued a report arguing that we were poised at a point of great scientific and technological opportunity. Science had gained a much greater appreciation of the degree to which controlling the structure of matter on the nanoscale could determine the macroscopic properties of materials. Further, major advances had been made in our ability to manipulate matter at the atomic and molecular levels. The working group report recommended the initiation of a national program in nanotechnology, which it defined as "the creation and utilization of materials, devices, and systems through the control of matter on the nanometer-length scale, that is, at the level of atoms, molecules and supramolecular structures." Nanotechnology, it was said, "could impact the production of virtually every human-made object . . . and lead to the invention of objects yet to be imagined."[5]

In July 2000, a national initiative in nanotechnology was proposed at a funding level of $225 million, an 83 percent increase in U.S. federally funded nanotechnology research in FY 2001. Congress approved a $152 million (+56 percent) increase in funding to support what is now known as the National Nanotechnology Initiative (NNI),[6] bringing federal funding for nanotechnology to $422 million in FY 2001. U.S. federal spending for 2002 is currently estimated at $600 million, an increase of an additional 42 percent.

The NNI called for a coordination of nanotechnology efforts at the Department of Defense, the Department of Energy, the National Science Foundation, the National Institutes of Health, the National Institute of Standards and Technology, and the National Aeronautics and Space Administration. It established a series of Grand Challenges, whose success would make a substantial difference in the lives of the U.S. citizenry.

While nanotechnology is in its infancy and substantial changes in direction and emphasis can be expected in the research portfolio of the NNI over the next decade, an inspection of the current NNI Grand Challenges gives an overview of the initial intent. There are nine NNI Grand Challenges in all, which are briefly described below.

- *Nanostructured materials by design.* By understanding the effect of a material's nanostructure on its macroscopic properties and by developing new methods of fabrication and measurement, produce materials that are stronger, lighter, harder, self-repairing, smarter, and safer.
- *Nanoelectronics, optoelectronics, and magnetics.* Develop new device concepts and methods of fabrication and processing. These new devices will be integrated into existing systems, and new architectures will be developed. Modeling and simulation of complex systems over a broad

range of lengths will be used. Explore the novel optical properties available with photonic bandgap structures.
- *Advanced health care, therapeutics, and diagnosis.* Improve our health by developing new medical imaging techniques and biosensors; use nanoparticles to target the delivery of drugs directly to the most important sites; improve biological implants through nanoscale engineering of the implant-bone interface; develop devices to enable vision and hearing; produce better diagnostic methods using gene sequencing methods.
- *Nanoscale processes for environmental improvement.* Develop a fundamental understanding of nanoscale processes important to maintaining environmental quality and controlling and minimizing unwanted emissions. Find methods to both measure pollutants and remove nanoscale pollutants from the water and air.
- *Efficient energy conversion and storage.* Obtain more efficient sources of energy through improved nanocrystal catalysts, more efficient and color tunable solar cells, and efficient photoactive materials for solar conversion of molecules to fuels. Explore the potential of carbon nanotubes for high-density hydrogen storage and the use of fluids with suspended nanocrystalline particles to improve the efficiency of heat exchangers. Develop high-performance magnetic materials and devices using nanoscale, layered materials. Develop high-efficiency light sources using layered quantum well structures.
- *Microcraft space exploration and industrialization.* Scale down the size of a spacecraft by an order of magnitude, allowing continuous exploration of space beyond the solar system. Utilize the high strength and light weight of nanostructured materials to greatly reduce fuel consumption, as well as permit the construction of large systems using inflatable structures, e.g., antennas. Enable sophisticated autonomous decision making through the development of nanoelectronics and nanomagnetics and the ability to store large amounts of data. Use self-repairing materials to further extend the reach of our exploration. Develop MEMS to allow for efficient control of optical surfaces and extend the development of radiation-tolerant technologies.
- *Bionanosensor devices for communicable disease and biological threat detection.* Greatly improve both the detection of and response to threats from chemical or biological warfare and from human disease through advances in nanoscience and nanotechnology. Improve health and extend human capabilities using nanodevices. In vivo nanodevices will derive power from body chemistry in the same ways natural molecular motors function and will be able to detect disease or chemical or biological threats. These nanodevices will also be capable of counteracting the threats of bioterrorism and chemical warfare. Perform critical research on issues relating to the compatibility between nanodevice materials and

living tissue. Microfluidics and sensor fouling will be important research areas.

- *Application to economical and safe transportation.* Develop more efficient means of transportation based on nanomaterials to provide lighter materials with lower failure rates; more durable materials for roads and bridges; smart materials that detect impending failure and carry out self-repair; low-friction and low-corrosion coatings; performance and emissions sensors; and emissions traps.
- *National security.* Use nanoscience and technology to assure military dominance at reduced cost and manpower, yet with lower risk to war fighters. Produce enormous gains in knowledge superiority with orders of magnitude gains in processor speed, storage capacity, access speed, display technology, and communications bandwidth. This will, for example, result in near-instantaneous worldwide communication, improved threat identification, communications encryption, extended multispectral imaging, and improved training through virtual reality. Military platforms will benefit from new nanomaterials through the development of self-assembly methods to reduce manufacturing costs and materials with improved properties, e.g., higher strength-to-weight ratio and low multispectral visibility. Nanodevices will protect the war fighter through detection of chemical and biological hazards and extend the war fighter's capability by extending his senses.

WORLDWIDE PERSPECTIVE

The increase in interest and investment in nanotechnology has been a worldwide phenomenon. Mirroring the growth in interest in the United States in the late 1990s, several studies of nanotechnology were done outside the United States. As the century ended, worldwide investment in nanotechnology was increasing. Figure 2-4 shows the rapid acceleration in spending by governments on nanotechnology research and development from 1997 to 2001. The increase amounted to 364 percent for the United States, which approximates the rate of increase worldwide.[7]

Most national programs cover science broadly; some have been designed to support the development of specific industries. The training of future scientists in what is perceived to be a key technology of the 21st century is generally an important part of each national program. The programs also stress the development of new manufacturing methods to derive the full benefit from nanoscience. A great deal of effort is going into collaborative efforts in this multidisciplinary field. National and international collaborations take the form of agreements between groups in the same discipline, between groups in different disciplines, or between groups in academe, industrial laboratories, and national laboratories. A variety of networks have been formed to bring groups together to share expertise and specialized equipment.

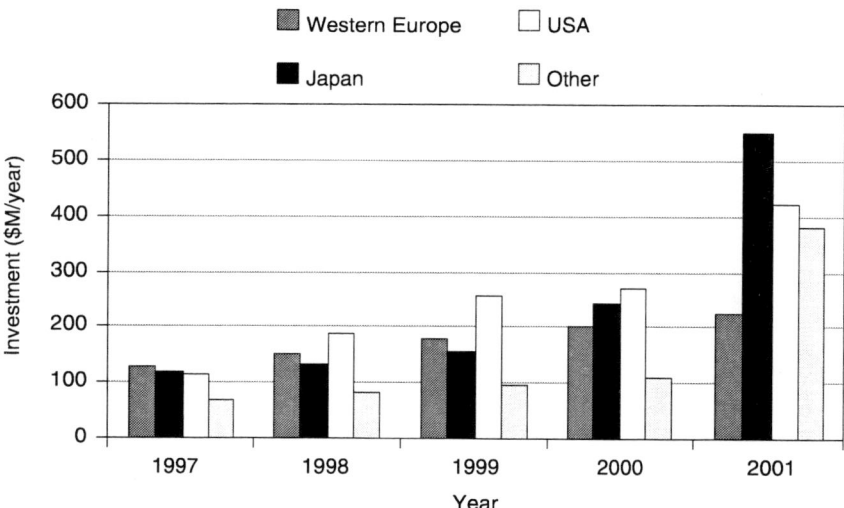

FIGURE 2-4 Worldwide government R&D spending on nanotechnology. Note that Western Europe includes the European Union and Switzerland. Countries included in the "Other" category are Australia, Canada, China, the former Soviet Union, Korea, Singapore, Taiwan, and additional countries. Finally, the entry for Japan in 2001 includes a $140 million addition to the budget that may include some programs outside the definition of nanotechnology used by the NNI. SOURCE: Data from M.C. Roco, 2001. International strategy for nanotechnology research and development. Journal of Nanoparticle Research 3(5-6): 353–360.

REFERENCES

1. International Technology Roadmap for Semiconductors. 2001. Available online at <http://public.itrs.net/> [July 1, 2002].
2. International Technology Roadmap for Semiconductors. 2001. Available online at <http://public.itrs.net/> [July 1, 2002].
3. Compañó, R. 2000. Technology Roadmap for Nanoelectronics, second edition, November. Luxembourg: Office for Official Publications of the European Communities.
4. MEMS Industry Group. 2001. 2001 Annual Report. Available online at <http://www.memsindustrygroup.org/arord01.htm> [July 2, 2002].
5. Roco, M.C., R.S. Williams, and P. Alivisatos. 2000. Nanotechnology Research Directions, IWGN Workshop Report. Boston, Mass.: Kluwer Academic Publishers.
6. National Science and Technology Council. 2000. National Technology Initiative: The Initiative and Its Implementation Plan. Available online at <http://www.nano.gov/nni2.pdf> [July 2, 2002].
7. Roco, M.C. 2001. International strategy for nanotechnology research and development. Journal of Nanoparticle Research 3(5-6): 353–360.

3

Major Areas of Opportunity

INFORMATION TECHNOLOGY

Introduction

The dramatic improvements in information technology over the last 50 years have led to a revolutionary change in the conduct of warfare. Information superiority is identified by the Air Force as a Core Competency, critical to modern warfare. The Critical Future Capability statement demands that "continuous, tailored information be provided within minutes of tasking with sufficient accuracy to engage any target in any battle space worldwide."[1] Information superiority is a critical component of other Core Competencies, including Aerospace Superiority, Global Attack, and Precision Engagement. Each of these is critically dependent on accurate and timely information.

Several pieces must come together to satisfy the Air Force's requirements. Sensors, discussed in the next section, provide the raw data. Electronic signal processing is applied to the sensor outputs to interface with the larger-scale information processing and communication systems. Communication at many levels is necessary to gather the information. Information processing, including fusion of data from multiple sensors, distills the sensor data into the information necessary for decision making. At each step there are requirements for data storage and display as well as for computation. Ultimately this must be a robust, redundant system tolerant of the failure of individual segments and self-reconfigurable to adjust to changing conditions and demands.

Research and development (R&D) investments by the Air Force in these areas must be considered in context with other worldwide efforts, especially those of industry. Because the DoD is today a relatively small customer for information technology, industrial R&D programs are considerably larger than those affordable by the Air Force. Clearly, the commercial sector provides immense incentive for going after scientific and technological advances in this field. However, some technologies that are unique to the military or that are not yet commercially viable require military investment:

> ... For example, many sensor applications are unique to government requirements and hence are funded solely by the government. Similarly, there are additional technologies that are essential for government missions but which may have or develop commercial application as well; however, the cost of their development is usually so high that industry cannot make a business case for maturing them commercially. Examples include the Global Positioning System, or development of new propulsion concepts.[2]

Assessing the appropriate R&D investment by the Air Force in light of the ongoing revolution in information technology in the commercial sector is challenging. In this section the committee examines some of the specific sectors of information technology (IT) to draw distinctions for the Air Force.

This section starts with computing devices. Transistors, switches, and integrated circuits are covered; a separate section on space electronics is included because of the unique environmental requirements of space and its importance to the Air Force. Also covered are storage and display technologies. Computing architectures explores alternative paradigms for computing. Under communications, a number of areas are explored: optical materials and devices, radio frequency (RF) materials and devices; and RF and optical MEMS. Finally, information and signal processing and data fusion requirements are discussed.

Computing Capabilities—Devices

Scaled Complementary Metal Oxide Semiconductors

Advances in information technology are a result of the ever-shrinking transistor, applied to almost every aspect of gathering and treating information. Continuing increases in information technology capabilities are dependent on continuing advances in the fabrication of ever more powerful computational hardware. Moore's law, the exponential increase in integrated circuit functionality, continues today. Although there are potential limitations on the horizon, the semiconductor industry roadmap, ITRS,[3] which is based on the scaling from larger devices that has served us so well for the last 40 years, foresees a continuation of the current rate of Moore's law to the present roadmap horizon of 2016, corresponding to a 32-fold improvement in device density. Barriers to reaching and surpassing this density include the following, among others: lithography, gate oxide current leakage, interconnect requirements, and thermal issues. These challenges have spurred many research efforts, both to address the issues within the context of traditional scaled-silicon systems and to find alternatives that circumvent the approaching barriers. Of great interest has been the field of nanotechnology, where multiple materials are innovatively positioned with nanometer precision.

The ITRS, discussed in Chapter 2, is industry's best analysis of all factors (fabrication, interconnects, thermal management, cross talk, cost, packaging, etc.) that must be considered in order to continue the miniaturization progress. The ITRS identifies a number of "brick walls"—major technology issues for which there is no known solution—as well other important technological challenges for which promising approaches have been identified. Given the current state of knowledge of CMOS and of the alternatives as they are now understood, the best guess for the next 10-15 years is that silicon CMOS technology will continue to provide the fastest switching time at the lowest cost in the smallest gate with the most cost-effective system integration.

The continuing hegemony of CMOS devices and circuits is based on substantial improvements and the introduction of highly innovative ideas. At the 2001 International Electron Devices Meeting, Intel announced a transistor operating at 3.3 terahertz.[4] At this same meeting, Advanced Micro Devices (AMD) announced that variations on CMOS transistors operate with 15-nm feature sizes. A November 26, 2001, announcement by Intel disclosed a "depleted substrate transistor" having a leakage current 100 times smaller than present transistors and, therefore, a 10^4 smaller gate leakage power.[5] This innovation could contribute substantially to the alleviation of heat dissipation that currently looms as a major issue. Other conventional approaches promote the use of asynchronous design or self-timed circuits operating without a single, chipwide clock speed orchestrating the tempo of each transistor.[6] CMOS and its many variations represent opportunities for vast improvements as nanoscale dimensions are reached.

Nonetheless, there are ultimate barriers to continued CMOS scaling, and new approaches for new devices and functions are being explored. Quantum interference effects, for example, may provide opportunities for new devices and functions. Some of the new approaches are based on alternative designs for transistors, while others represent entirely new ideas for logic operations. It is clear that the current architecture for digital computers is not unique, nor does it provide the greatest capability for some operations. The brain is able to process information for operations such as image recognition with far greater speed and efficacy than current computational approaches. Just how alternative architectures may operate, and which ones are likely to provide substantial improvement for certain operations, remains a frontier of current research.

There have been many attempts to think outside the box. Radically different approaches are being investigated that, in most cases, attack only one small element of what, ultimately, must be an integrated effort tying together many factors that must be satisfied simultaneously for such a system to be of practical use (in a manner similar to the ITRS). These new approaches, some of which stem from frontier scientific discoveries, illuminate aspects of the difficult terrain ahead. Reviews of many of these approaches and observations on their performance are available.[7,8,9]

Enthusiasm for fabricating logic circuits from molecules is currently high. Tour-de-force feats with carbon nanotubes (see next section) and molecular-layer transistors have succeeded in actually fabricating transistors and even simple logic circuits. The progress made in 2001 was recognized as the breakthrough of the year by the magazine *Science*.[10]

With such significant progress, it may appear that technological developments are imminent. However, there appears to be no single approach with a clear path to competitive products. Integration of these individual elements into a highly dense fabric with functionality approaching today's integrated circuits remains a formidable challenge, particularly in the face of our relatively rudimentary fabrication capabilities at the nanoscale (see Chapter 4). A number of approaches are being pursued with the goal of revealing additional scientific information that might improve prospects. These limitations have been discussed in articles by Meindl[11,12,13] and Thompson, Packan and Bohr.[14] The challenge presented by the appropriate design of interconnects has also been investigated extensively.[15,16,17,18] In this context, it is important to acknowledge the extreme sophistication of the current integrated circuit paradigm and to recognize that the most likely early adoptions of these new technologies will be as adjuncts to, rather than replacements for, the manufacturing technology that currently dominates the marketplace.

A recent paper attempts to introduce some logic to the plethora of current research directions.[19] Of great importance is the observation that the transistor, the basis for such tremendous gains in information technology, has features that are critically important to its success: (1) it offers high gain, which allows a single transistor to reset logic levels after each stage and to drive multiple following transistors (fan-out) and (2) it isolates the output of the device from the input. These features allow the signal for a bit, with inevitably irregular amplitudes and features, to be combined with signals for other bits, yielding reliable logical functions and the accumulation of a result that maintains its integrity in a noisy environment. Other devices, such as resonant tunneling diodes (RTDs), can be used to generate gain; but because they are two terminal diode devices, they lack the isolation required for robust accumulation of logic operations. The advantages of RTDs with respect to switching speed have been known for years, but no uses for these devices have been found for logical operations in computers.

The author of the above-mentioned paper, Keyes, concludes with the following paragraph:

The fact that transistors have had no competitor in digital electronics for 40 years does not imply that no alternatives should be sought and studied. However, a search for a new concept must include an awareness of what digital means:

a well-defined value for a digit and a way of maintaining and setting a signal to that value in a noisy environment, with mass-produced imprecise components. Alternatively it should be perfectly clear that digital representation is being abandoned if that is indeed the case, and that there is another way to cope with the inevitable uncertainty in the parameters of devices and the distortion of signals propagated in a large system.

With this caveat in mind, the committee surveyed the various technologies that are currently under investigation as possible successors to the silicon CMOS mantle.

Single-Electron Transistors

The operation of a single-electron device, discussed in detail in a recent review,[20] is based on the fact that the charging energy to add an additional electron to a small island, generally through a tunneling barrier, becomes significant if the island is of nanoscale dimensions. Initial research was performed at low temperatures, where thermal fluctuations remain negligible for nanostructures in a readily achievable size range. However, fabrication of structures in the 1-nm range will allow stable room temperature operation.

Likharev points out two major unsolved challenges that face developers of single-electron transistor logic.[21] First, there is the deleterious effect of random stray charges embedded in nearby insulators. These stray charges produce random, time-varying background charge levels, which impact device thresholds.[22] Second, subnanometer structures will be needed at the heart of the single-electron device to allow room temperature operation. They will need to be very regular in size and shape to assure uniform device performance. If self-assembly fabrication methods (see Chapter 4) are used to generate perfectly regular nanostructures, they must necessarily be incorporated into a larger microstructure. The precise placement and interconnection of these subnanometer structures, including the placement of suitable tunneling barriers, present a formidable challenge to schemes for nanoassembly. Since these devices operate at the level of individual electrons, there is a fundamental issue with gain and fan-out (the ability to drive multiple following stages from a single device output) that poses a significant architectural problem for large systems.

Spin-Based Electronics

Following the development of the very successful giant magnetoresistive read heads for magnetic storage and magnetic field sensors, a new area of spin-based electronics is emerging.[23] The concept is to use the spin of the electron in suitably designed devices to perform logic operations. One can imagine that information now stored as the presence or absence of charge could alternatively be stored as spin-up or spin-down. This might be particularly attractive if the spin information can be communicated from one point in the circuit to another without moving the corresponding charges, which would eliminate the accompanying dissipation mechanisms. This research is in its initial phases, asking basic questions about the distances over which spin can be transported without depolarization, the characteristics of sources and detectors of electron polarization, the effects of interfaces between ferromagnets and semiconductors, and the feasibility of ferromagnetic semiconductors in such applications.

Domains in ferromagnets, which are formed by spins coupled together by the exchange interaction, are also being explored for use in computing. The propagation of the orientation direction of a magnetic domain through a series of nanoscale dots has been recently demonstrated.[24] Further, using domain orientation to represent logic states, it has been possible to experimentally demonstrate the functionality of both a NOT gate and a shift register using continuous nanoscale magnetic "wires."[25]

Molecular Electronics

In its present incarnation, the term "molecular electronics" was coined shortly after the discovery of conducting organic polymers in the late 1970s in recognition of the significant electrical conduction properties of many organic materials. A second driving factor was the clear recognition that the brain of a living species represented a logical device with the ability to recognize an image much more rapidly that digital computers. This was the proof of theorem for a "molecular computer," and the vision grew. Today molecular electronics has come to mean the use of molecules in electronic devices.

Initial visions of molecular electronics focused on how the arrangement of chemical bonds in these molecules might function as circuits and switches. A significant number of researchers are exploring this area, although even after two decades it has not been possible to experimentally verify many of the initial hypotheses. Some direct measurements of electrical conductivity across single molecules have verified the magnitude of the electrical conduction found in single-molecule layers and illuminated possible mechanisms for conductivity in bulk molecular materials. Many of the experimental observations involving molecular conductivity behavior are puzzling and have not been explained fully.

As the behavior of molecular units becomes better understood, it must be emphasized that a wide range of issues faces their practical application. The many considerations and challenges in the ITRS indicate the complexity of designing, fabricating, testing, and packaging chips with 0.13-micrometer feature sizes today. All of these issues are likely to be considerably more complex as dimensions are further reduced. This suggests that the first uses of molecular electronics are likely to be as adjuncts to, rather than replacements for, the integrated circuit.

As activity in molecular electronics expanded, it was recognized that molecular materials might provide significant advantages for many other electronics applications. Organic thin-film transistors (the bulk organic material) demonstrate significant gain with reasonable characteristics for some electronic devices.[26,27] The mobility of charge carriers in these materials is about three orders of magnitude lower than for commonly used semiconductors. This is fundamentally due to the hopping mechanism for conduction in disordered materials (as compared with band conduction in crystalline materials, which relegates such devices to speeds much slower than those currently achieved with today's CMOS devices). But organic transistors fabricated by a variety of methods offer advantages such as low temperature, low-cost formation of large-area arrays, with particular potential for applications involving flexible structures (e.g., products such as credit cards and displays).[28] Ink jet printers have been used to achieve transistor gate lengths of 5 micrometers and also to fabricate arrays of organic light-emitting diodes.[29] Ink jet techniques have been also extended to such unconventional areas as deposition of suspended alloys and metallic or magnetic nanoparticles offering advantages for electronic applications.[30] Organic transistors are envisioned for use as switching devices for active matrix flat panel displays (liquid crystal, organic light-emitting diodes, and "electronic paper").[31] In addition, organic and semiconductor white-light-emitting structures are anticipated to come into use in the future and to have a significant impact on energy use. All-polymer integrated circuits for use as radio frequency identification tags and for various sensors have been proposed. A variety of materials and methods are under examination with the purpose of developing low-cost, continuous-feed or reel-to-reel production methods for these low-end applications.

Carbon Nanotube Electronics

Carbon nanotubes (CNTs) are a unique material (see Box 3-1) with remarkable electronic, mechanical, and chemical properties. CNT electrical behavior is different from that of ordinary conductors. Depending on the application, the difference in behavior might be an advantage, a disadvantage, or an opportunity. Electrical conduction within a perfect nanotube is ballistic, with low thermal dissipation, an advantage for computer chips if the tubes can be seamlessly interconnected. Perturbations such as electrical connections modify this behavior substantially. For slower signal speed for analog processing, nanotubes surrounding buckyball molecules acting as transmitters have been experimentally demonstrated. Individual multiwalled CNTs at room temperature exhibit quantized conductance at values of $G = 2e^2/h$ (= 12.9 kΩ^{-1}), a remarkable observation. The current density for these experiments was 10^7 A/cm^2, a value two orders of magnitude greater than current densities normally available for superconductors.

Field-effect molecular transistors[32] have been demonstrated using a back gate with a carbon nanotube[33] settled across two gold conductors. The challenge

BOX 3.1
The Ubiquitous Carbon Nanotube

Discovered in 1991, tubular structures of carbon had been predicted since the discovery of soccer-ball-shaped 60-carbon molecules (buckminsterfullerenes, or "buckyballs") in 1985 at Rice University. Each carbon atom in a nanotube is positioned in a lattice that wraps into a hollow pipe ranging from a few to tens of nanometers in diameter. Figure 3-1-1 shows various carbon nanotube structures, including multiwalled and metal-atom-filled nanotubes.

Because of their unique self-assembled and atomically perfect structures, carbon nanotubes exhibit unusual electrical, mechanical, and chemical properties. These special properties, such as the ability to carry exceptionally high current densities in long molecularly perfect "wires" and unusually high mechanical strength at the limit of small 'fiber' diameters, have generated much interest in the potential applications of nanotubes.

Displays

Depending on their diameter and chirality, nanotubes exhibit either metallic (like copper) or semiconducting behavior. Metallic nanotubes can emit electrons from their extremely fine tips at quite low-voltages. The possibility of fabricating

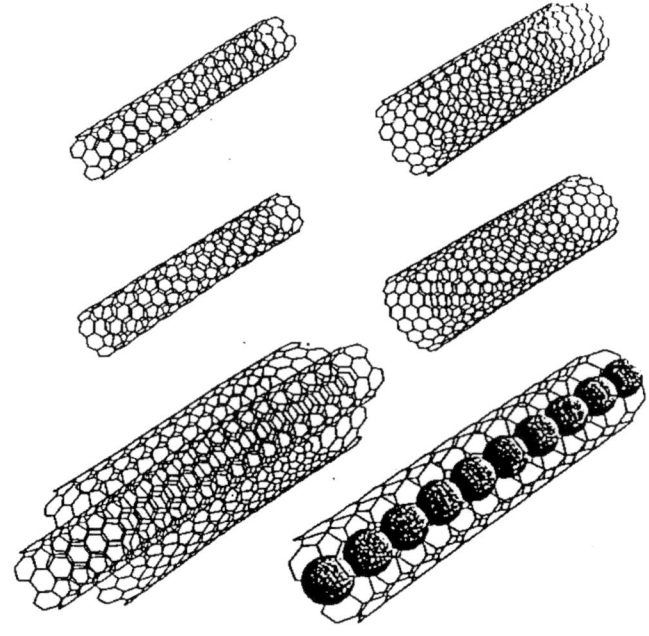

FIGURE 3-1-1 Carbon nanotube structures.

continues

nanotube arrays on surfaces for efficient current emitting elements is of great interest for low-power field emission displays, and a number of companies are racing to develop flat panel displays for next generation television and computer screens.

Computing

Further down the road, computer memory and logic concepts based on carbon nanotubes are being explored. Transistors made from carbon nanotubes a few nanometers in diameter—a hundred times smaller than the 130-nanometer transistor gates now found on computer chips—have been demonstrated.[1] Also demonstrated are collections of nanotube transistors working together as simple logic gates, the fundamental computer component that transforms electrical signals into meaningful ones and zeros. If nanotubes or related nanowires could be used as tiny electronic switches or transistors, computer designers could, in principle, cram billions of devices onto a chip (the Pentium 4 has only 55 million transistors).[2] The real challenge in this or any other alternative nanotechnology approach to fabricating computer chips is to design and connect up many millions or billions of such components in a highly manufacturable and reliable architecture.

Mechanical Properties

As a result of their seamless cylindrical structure, carbon nanotubes have low density, high stiffness, and high axial strength. Theoretical studies and recent experimental measurements suggest that the Young's modulus and breaking strength of single-wall carbon nanotubes are exceptionally high.[3] Carbon fibers with a tensile strength up to 6 GPa are commercially available, while initial experimental measurements on 4-mm-long single-wall carbon nanotube (SWNT) "ropes" consisting of tens to hundreds of individual SWNTs have yielded values up to 45 GPa.[4] The hope is that millimeter-long SWNTs can be formed into longer fibers or dispersed into a composite matrix while still maintaining a significant fraction of this observed improvement over conventional carbon fibers. The major challenge is retaining the strength of nanofibers and assemblies of nanofibers in conjunction with a matrix material so these properties can be controlled, optimized, and made practical.

Energy Storage

Carbon nanotubes could be used to improve batteries. They can in principle store twice as much energy density as graphite, the form of carbon currently used as an electrode in many rechargeable lithium batteries. Conventional graphite electrodes can reversibly store one lithium ion for every six carbon atoms. Tiny straws of carbon tubes reversibly store one charged ion for every three carbon atoms, double the capacity of graphite.[5] Carbon nanotubes are also being investigated for hydrogen storage. They may be capable of storing amounts comparable to or exceeding the U.S. Department of Energy target of 6.5 percent of their own weight in hydrogen, a level considered necessary to be practical for fuel cell electric vehicles.[6]

The carbon nanotube is a now-classic example of a well-defined nanostructure, and exploring ways to exploit its unique properties for possible nanotechnology-based applications remains a subject of intense interest. A general issue is the ability to reproducibly obtain large quantities of selective configurations of single- or multiple-wall nanotubes. Methods for synthesizing nanotubes, controlling orientation, and producing macroscopic quantities of these nanostructures are advancing but still are at an early stage of development. Integrating these materials into, for example, a composite matrix or an interconnected electrical structure, where their nanoscale properties translate into macroscale effects, remains a key challenge.

---

[1] See references in Service, R.F. 2001. Molecules get wired. Science 294(5551): 2442–2443 for a review of the current status of carbon nanotube electronic devices and circuits.
[2] Wasson, S., and A. Brown. 2002. Pentium 4 "Northwood" 2.2 GHz vs. Athlon XP2000+ Battle of the big dawgs, January 7. Available online at <http://www.tech-report.com/reviews/2002q1/northwood-vs-2000/index.x?pg=1>[April 24, 2002].
[3] Yu, M-F, B.S. Files, S. Arepalli, and R.S. Ruoff. 2000. Tensile loading of ropes of single wall carbon nanotubes and their mechanical properties. Physical Review Letters 84(24): 5552–5555.
[4] See references in Service, R.F. 2001. Molecules get wired. Science 294(5551): 2442–2443 for a review of the current status of carbon nanotube electronic devices and circuits.
[5] Scientific American. 2002. Carbon nanotubes could lengthen battery life. Scientific American News in Brief, January 9. Available online at <http://www.sciam.com/news/010902/2.html> [April 24, 2002].
[6] Dagani, R. 2002. Tempest in a tiny tube. Chemical and Engineering News 80(2): 25–28.

---

of demonstrating logic circuits was met recently with electrostatically doped CNTs. Doping of CNTs may be accomplished chemically;[34] CMOS-type inverters have been demonstrated with both p- and n-type doping. This work experimentally demonstrated the performance of an inverter and a NOR circuit using transistor-resistor logic with an on-off ratio of 10^5 and high gain.[35] It is clear that by arranging these CNT transistors appropriately, the functions AND, OR, NAND, and XOR can be realized. While these results represent a tremendous achievement, many questions remain.

> Even if all goes well, most experts predict it will be at least a decade before nanotubes become a significant part of computers. Challenging the supremacy of silicon is an enormous technical and financial task that will take far more than some promising scientific advances. It will take equally impressive advances in manufacturing and computer design. "Nanotubes can be used as transistors, logic and memory; all that has been demonstrated now," says Hongjie Dai, a chemist and nanotube researcher at Stanford University. "The question now is, how practical can these [nanotube] devices be?"[36]

Quantum Interference Devices

The term "quantum devices" refers to devices dominated by nonclassical effects arising from the discrete nature of matter at atomic dimensions and the resulting wave interference effects. The RTD is a device exhibiting negative differential resistance (NDR) owing to interference effects (or, equivalently, owing to resonant energy transfer across quantum levels). It is already well accepted as a device capable of enhancing the speed of field-effect transistor logic devices by factors of 2 to 5, allowing a significant increase in processing speed for digital signal processors (DSPs).[37]

Conductance through a medium having lateral dimensions smaller than an electron wavelength is quantized in units of $2e^2/h$, where e is the charge on the electron and h is Planck's constant,[38] as a consequence of the discrete energy levels along with the Fermi velocity and electron density of states at the Fermi level. Geometric imperfections and impurities in the conduction medium give rise to many kinds of variations in the conduction behavior of nanostructures.

Interference effects have been the subject of extensive research (see, for example, Agranovich).[39] The Aharonov-Bohm effect occurs when charge carriers are passed through a ring, where two separate paths of almost equal length are possible for these carriers. As they meet at the other side of the ring, if the distance traversed is less than the coherence length, interference effects appear as a function of magnetic field. This magnetoresistive behavior is fairly straightforward for regular geometric features. In more irregular shapes, however, the prediction becomes increasingly difficult owing to the complexity of the wave equation solutions in the presence of more complex boundary conditions. Magnetoresistive measurements of most objects with dimensions smaller than a coherence length demonstrate these "conductance fluctuations," which have been examined extensively. This serves to alert investigators to the sensitivity such nanostructured devices are likely to have to the presence of geometric irregularities. These wave effects form the basis of quantum computing approaches (see section on computing architectures) that might vastly increase the ability to solve certain important classes of problems, such as prime number factorization.

Solid Electrochemical Switching

A new method of switching using a nanoscale device was recently described.[40] A tip made of a solid electrolyte, silver sulfide, is positioned a few nanometers above a flat platinum surface. When a bias voltage as low as 10 mV is applied across the gap, atoms come out of solution and extend the tip toward the surface, eventually making contact. Quantized conductance is observed with conductance values of $n(2e^2/h)$, where $n = 0$ through 5, depending on the voltage applied. It is expected that this reversible process can be controlled with switching rates of 100 megahertz and on-off impedance ratios of 1:1,000 in air at room temperature. Simple logic gates have been constructed, which may have applications in information storage.

Vacuum Microelectronics: Back to the Future

Before transistors, vacuum tubes provided gain for electronic circuits. Today, vacuum tubes are still used as high-power radio frequency generators and amplifiers and as display devices (most televisions and computer monitors are still based on cathode ray tubes). Vacuum tubes utilize the free-space transmission of electrons from cathode to anode and are inherently radiation-hard and capable of handling kilowatts to megawatts of power. Figure 3.1 shows historical power and frequency limits for various vacuum RF generators.[41] With the exception of the free-electron laser (FEL), these devices rely on space-charge waves. As frequencies increase beyond 300 gigahertz and free-space wavelengths drop below 1 mm, characteristic lengths for modulation elements will drop below hundreds of micrometers. Micromachining of some type will be required. Novel high-power millimeter and submillimeter wave sources may be possible through batch fabrication of parallel oscillators or amplifiers with appropriate phase-locking techniques.

In traditional vacuum electronic devices, free electrons are extracted from thermionic cathodes by applying an electric field. These cathodes must operate at temperatures in excess of 700°C, thus requiring a heat source and a thermally isolated mounting. Vacuum tubes and their cathodes have limited lifetimes; vacuum tube-based computers in the 1950s were limited in their complexity by the failure rate of individual tubes.

Micro- and nanotechnology offer more robust cathodes, based on field emission, that can operate at room temperature and below. This enables new vacuum electron devices such as flat cathode ray tubes, smaller and more robust x-ray tubes for inspection and sterilization, and vacuum microelectronic integrated circuits for extreme temperature and radiation environments. Vacuum microelectronic flat panel displays utilize tens to thousands of active emitters per pixel,

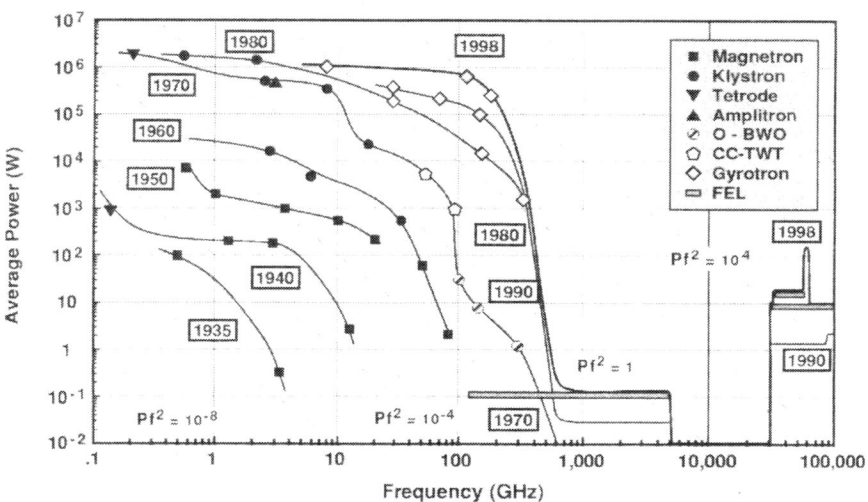

FIGURE 3-1 Power versus frequency for high-frequency microwave devices. SOURCE: Granatstein, V.L., R. Parker, and C.M. Armstrong. 1999. Vacuum electronics at the dawn of the twenty-first century. Proceedings of the IEEE 87(5): 702–716. © 1999 IEEE.

operating in parallel, instead of the conventional one (monochrome displays) or three (color displays) electron sources that are used to irradiate all of the pixels in traditional cathode ray tubes. Each pixel has its own electron source, eliminating the need for electron beam steering and an associated beam "lever arm" (deflected flight path). The net result is a micrometer- to millimeter-long electron flight path (instead of tens of centimeters) and flat vacuum tubes. The shorter flight paths minimize space-charge effects in the electron beamlets, and the parallel irradiation of pixels reduces the required current per electron beam. Overall, this translates into electron beam energies significantly below a kilovolt, with corresponding increases in operational safety and power supply simplicity. More detail on the performance of these cathodes is presented in the section "Aerodynamics, Propulsion, and Power," along with a discussion of their application to propulsion.

Space Electronics

The Air Force's space mission has unique electronics requirements that are not a concern for the dominant terrestrial industry base. This means that the Air Force, along with other appropriate mission agencies, must sustain a long-term research program for adapting advances in electronics and computing to the space environment.

Man-made satellites orbit Earth at altitudes from 200 kilometers to more than 36,000 kilometers. The region from 200 kilometers to 1,500 kilometers altitude is called low Earth orbit (LEO), and in this regime Earth's atmosphere is still present, but at exceedingly low densities. Earth's atmosphere and magnetic field together protect terrestrial dwellers from the cosmic rays, high-energy protons, and high-energy electrons normally present in the space environment. High-energy particle radiation is many orders of magnitude greater on-orbit than on Earth's surface. Commercial CMOS electronics are designed to operate in our low-radiation biosphere (roughly 0.3 rad/year, where a rad is the amount of particle radiation that deposits 100 ergs of energy per gram of target material), but they can usually tolerate total radiation doses as high as 10 kilorads.

Particle radiation damages electronic circuits; it creates single-event upsets (SEUs), latchups, and a gradual change in semiconductor current vs. voltage characteristics. The first effect causes erratic operation, the second effect can cause instant device destruction, and the last effect causes an inexorable increase in power consumption until the device ceases to function. SEUs are generated by a particle-induced transfer of charge to or from an active device. SEUs are minimized by increasing gate size and power requirements, by spot shielding of sensitive components, by using error detection and control (EDAC) circuitry or software, or by using silicon-on-insulator designs that decouple the substrate from the active regions.

MAJOR AREAS OF OPPORTUNITY 53

Latchup occurs when an ionizing trail creates a high-conductivity path between a current source and a current sink, generating a temporary electric short. Latchups are minimized by adding guard bands to transistors and by using silicon-on-insulator designs. The latter approach is used in fabricating radiation-hardened electronics for space applications. If latchup occurs, it may be corrected by momentarily removing power to the affected circuit. Latchup conditions are detected by monitoring power consumption to individual chips, or by adding "watchdog" timers.

Table 3-1 gives approximate radiation hardness levels for different types of semiconductor devices that were available in 1990. The actual radiation tolerance varies widely from design to design and is also fabrication-process-dependent, so radiation testing should be performed on selected components. Transistor-transistor logic (TTL) and emitter-coupled logic (ECL) circuits are inherently more radiation hard than CMOS, but they require more power. Metal-oxide (n-type) silicon (NMOS), metal-oxide (p-type) silicon (PMOS), current-current logic (I^2L), and silicon-on-insulator metal oxide semiconductor (MOS) circuits can be fully immune to latchup. CMOS circuitry fabricated onto silicon-on-insulator substrates has traditionally provided radiation-tolerant electronics for space applications. The use of thin silicon over an insulator reduces the volume for charge collection along an ionizing particle track, thus reducing the amount of charge introduced into individual gates. Thin-film silicon-on-insulator (TFSOI) technology is now being considered for commercial electronics because it can provide enhanced low-voltage operation, simplified circuit fabrication, and reduced circuit sizes relative to bulk silicon counterparts.[42] TFSOI would be particularly

TABLE 3-1 Approximate Radiation Hardness Levels for Semiconductor Devices

Technology	Total Dose (rads) (silicon)
CMOS (soft)	10^3-10^4
CMOS (hardened)	5×10^4-10^6
CMOS (silicon-on-sapphire: soft)	10^3-10^4
CMOS (silicon-on-sapphire: hardened)	$>10^5$
ECL	10^7
I^2L	10^5-4×10^6
Linear integrated circuits	5×10^3-10^7
MNOS	10^3-10^5
MNOS (hardened)	5×10^5-10^6
NMOS	7×10^2-7×10^3
PMOS	4×10^3-10^5
TTL/STTL	$>10^6$

SOURCE: Adapted from Griffin, M.D., and J.R. French. 1991. Space Vehicle Design. Washington, D.C.: American Institute of Aeronautics and Astronautics.

interesting for MEMS space applications because of its inherent radiation tolerance and its built-in etch stop for bulk silicon etching.

Can commercial electronics be used on-orbit with appropriate radiation shielding? Figure 3-2 shows the yearly total radiation dose in silicon electronics as a function of aluminum shield thickness for an 800-kilometer altitude Sun-synchronous orbit. Contributions due to trapped protons and electrons are shown; contributions due to solar protons and bremsstrahlung were not plotted for clarity. This graph was generated using data supplied by the on-line Space Environment Information System (SPENVIS).[43] Figure 3-2 indicates that commercial CMOS devices should have at least several millimeters of shielding to survive a year in LEO orbit.

Figure 3-3 shows radiation dose rate dependence as a function of circular equatorial orbit altitude for aluminum shielding with thicknesses of 0.18 and 1.1 centimeters (densities of 0.5 g/cm^2 and 3.0 g/cm^2, respectively). Note the rapid rise in dose rate with altitude above about 2,000 kilometers and below 20,000 kilometers. The hard-to-shield proton belt peaks at ~4,000 kilometers, and the easier-to-shield electron belt peaks at ~20,000 kilometers. At geostationary Earth orbit (GEO; 35,786 kilometers altitude and 0 degrees inclination) with a maximum dose of 3,000 rads, 0.5 g/cm^2 (0.22-centimeter silicon) and 3.0 g/cm^2 (1.3-

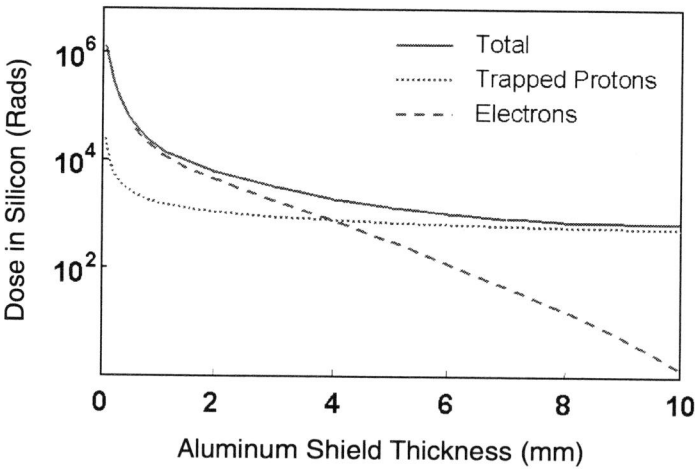

FIGURE 3-2 Yearly radiation dose in silicon. Adapted from data at European Space Agency, Space Environment Information System. Available online at <http://www.spenvis.oma.be/spenvis/> [April 24, 2002].

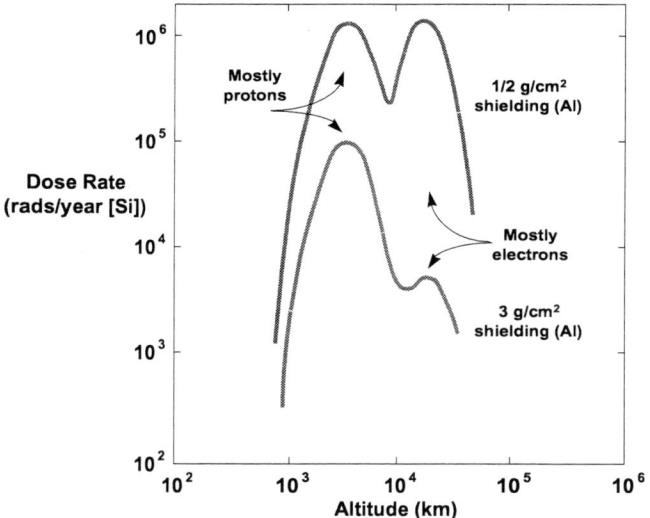

FIGURE 3-3 Radiation environment for circular equatorial orbits. SOURCE: Helvajian, H., and S.W. Janson. 1999. Microengineering space systems. Pp. 29–72 in Microengineering Aerospace Systems, H. Helvajian, ed. Reston, Va.: American Institute of Aeronautics and Astronautics, Inc. Figure reprinted with permission of the Aerospace Corporation.

centimeter silicon) shielding give lifetimes of roughly 11 days and 3 years, respectively. The significance of Figures 3-2 and 3-3 is that while CMOS circuitry can be used for low-altitude LEO missions, more radiation-resistant technologies may be necessary for other orbits.

Specialized radiation-hard devices are available, but they are expensive and are about two to three technology generations behind their commercial counterparts.[44] Fortunately, commercial CMOS foundry processes, in general, have increasing total dose hardness as device feature sizes decrease. The increased hardness apparently results from decreased gate oxide thickness.[45] One commercially available 0.25-micrometer process has an apparent total dose limit of greater than 100 kilorads without design changes and greater than 500 kilorads with the addition of guard bands, etc.[46] Latchup and SEUs must still be dealt with, however.

Finally, another back-to-the-future approach may prove useful for radiation-hard electronics. Fifty years ago, magnetic core memory composed of small (about 1/30 inch in diameter) ferrite toroids was used to provide non-volatile memory.[47] Information was stored as a magnetization state of the ferrite core, resulting in an inherently radiation-hard system that could also withstand high current transients. Variants of this technology were used to create simple logic circuits, which were integrated into B-52 bombers, Minuteman launch control systems, and some satellites. Today, giant magnetoresistance is being exploited by IBM, Motorola, and others to develop low-power nonvolatile memory, called magnetic RAM or MRAM (see the subsection "Magnetic Storage" in the section after next) that has the potential to replace Flash memory in consumer applications. Unlike Flash memory, magnetic memory would have an unlimited number of read/write cycles. Radiation hardness is not the driver for these applications, so the on-chip driver circuits will probably be standard CMOS. For radiation hardness, silicon-on-sapphire CMOS could be integrated with magnetic memory elements. A radiation-hardened all-magnetic solution would require replacing well-established CMOS circuits with less mature magnetic logic circuits. Near- to midterm devices will probably consist of CMOS/magnetic hybrids, while long-term devices could be all-magnetic if the technology proves practical.

The future of space micro- and nanoelectronics lies with a combination of commercial and radiation-hardened CMOS for digital and analog circuits. Radiation-hard fabrication houses must continue to be supported for non-LEO missions unless TFSOI technology becomes a commercial process. Vacuum integrated circuits are one possibility for radiation hardness over an extremely wide operating temperature range; this would be ideal for ultrasmall or essentially two-dimensional spacecraft with minimal mass for radiation shielding or thermal inertia. Total dose hardness of commercial CMOS may continue to increase with time, but latchup and SEU prevention will necessitate the use of specialized gate designs and additional on-chip protection circuitry. Scalable CMOS libraries with these attributes must continue to be supported.

For the mid- to far-term with even higher device densities and lower power operation, SEUs may become MEUs (multiple-event upsets) as particle radiation tracks impact more than one gate at a time. The charge required to change a digital one to a zero will ultimately drop to one electron. Spacecraft radiation tolerance will result from a combination of device (gate) hardness, on-chip mitigation techniques (EDAC circuitry, watchdogs, and fault-tolerant architectures), and system-level fault control (voting processors, field reprogrammable gate arrays, adaptive networks, and intelligent power control). As systems become more complex, more attention to fault detection and correction will be required.

Computing Capabilities—Architectures

Computing today is dominated by a von Neumann architecture with a central processing unit (CPU) as the focal point. Massively parallel architectures are becoming available, but generally with a coordination between processing units that derives from this centralized architecture. Two themes are resulting in the exploration of changes in this paradigm. The first is the new functions that may be possible, or even required, as the array of new devices being explored and discussed above becomes reality. The limits of interconnections and communications between devices are being reached with current architectures, driving an investigation of alternative schemes such as cellular automata. Quantum computing provides a tantalizing hint of orders-of-magnitude increases in computing power for some important operations such as factoring large numbers, but today it remains beyond reach and is the subject of intense worldwide investigation. The second theme is the evident superiority of the human brain over even the most advanced computers for some computational tasks such as image recognition. Biomimetic investigations have long sought to capture the processing of the brain, which involves analog functions and massive storage. In general it may be anticipated that alternative approaches to computing involving analog or other yet-to-be-invented methods may someday have an impact on computing.

Cellular Automata

The term "cellular automata" (CA) generally refers to a group or population of interacting cells in which each cell (automaton) behaves according to a well-defined rule set. Each cell has a limited number of well-defined states. The cells are dynamic, and the properties of each cell evolve with time in a manner dependent on the states of neighboring cells. Complex phenomena may be simulated with an appropriate set of criteria, much as fluid behavior "solves" the Navier-Stokes equation for interacting fluid "cells" or for population behavior by means of interactions among individual members of a group.

Quantum-dot cellular automata (QDCA) represent one such implementation of this basic idea. They consist of arrays of quantum dots in which the state of an array exists in various electronic configurations. The array evolves in time depending on the states of the interacting cells. A group at the University of Notre Dame has theoretically and experimentally investigated one such model extensively.[48,49] A recent review provides an excellent overview of this and related work.[50] The fundamental cell is a square array of four quantum dots (each dot at a corner of a square) and two electrons. The preferred locations of the two electrons are at opposite corners of the square. Since there are two ways to arrange these electrons, each having the same energy, this basic unit may represent a bit (one or zero) depending on the arrangement taken by the two electrons. By judicious placement of these cells, the behavior of wires and conventional logic gates may be mimicked. QDCA evolve with time in a continuous manner rather than in discrete time steps. Logical operations may be demonstrated; more complex behavior depends on the array geometry and size.

Input to a QDCA consists of initiating the state of several cells with neighboring charged wires. The circuit then relaxes with time to the lowest ground state. The output is read by measuring the polarities at output cells; this represents the solution to the problem (as defined by the geometry of the array). The behavior of these arrays has been simulated and demonstrated at the micrometer level with lithographically fabricated metallic islands, and logical operations have been demonstrated. Because of the large size of the dots, these arrays must be cooled to about 0.1 K to operate.[51] Smaller versions could operate at higher temperatures. Miniaturized versions of these arrays are envisioned to be suitably constructed molecules. One limit that appears for these systems is thermodynamic in nature. As the number of cells in the array increases to mimic more complex logical operations, the energy levels are necessarily closer together. If the energy level spacing approaches thermal energies (kT), the solution (the ground state) becomes mixed with higher energy states. By reducing the dimensions to about 2 nanometers, room-temperature operation appears feasible. This may be enabled by designing special-purpose molecules.[52] Alternatively, introducing multiple layers gives a greater separation of energy levels and is another approach to extending this behavior to room temperature. For the molecular implementation, leads, clocking arrangements, etc. must be introduced—all major challenges.

The use of coulombically coupled quantum-dot cellular arrays for quantum computing (see following section) has been suggested.[53] The various qubit operations may be stimulated by raising and lowering the interdot tunneling barriers with electric, magnetic, or optical fields. The nature of coherence for the overall system must be understood. A very significant issue is that the decoherence time for solid-state quantum media is significantly shorter than for isolated atomic or molecular systems.

An understanding of how to utilize QDCA effectively in solving a problem must be developed. QDCA are one approach to parallel device operations. In addition to providing a compact approach to complex logical operations, they may have widespread application in image processing and other highly parallelizable problems. Research is continuing to gain an understanding of the dynamic properties of these systems. It is clear that much needs to be learned before the feasibility of fabricating and implementing such devices, and their standing relative to competing architectures, can be understood.

Quantum Computing

Although quantum computation holds the promise of an exponential increase in computation speed and therefore application to important problems of greatly increased complexity, so far only a few specific algorithms that gain this advantage are known,[54] and hardware systems that are scalable to a sufficient number of qubits (quantum bits) for useful application have yet to be demonstrated. The primary problem is that quantum wave functions are notoriously fragile and susceptible to disturbances that restrict the coherence time of the system. The initial state must be definable, the quantum states must be uniquely addressable during execution of the algorithm, and the final state of the system must be observable. Additional algorithms must be discovered and a useful quantum computation must be demonstrated before thought can be given to practical applications.

The basic quantum computer consists of a loosely coupled array of binary quantum states, each of which represents a single qubit. For such a linear system, the ideal initial state consists of the ground state, which is then addressed by a series of unitary transformations in the form of energy pulses, addressing each qubit in a unique fashion in a sequence to execute the desired algorithm. This is followed by observation of the final state of the assembly to determine the "answer." The system must be coherent throughout the entire process. Quantum computers are in some sense more like analog computers than digital computers. Existing algorithms take advantage of hidden resonances in the problem, exploiting them to obtain the exponential increase in performance.

There is great interest in developing quantum computation. Numerous approaches are currently being investigated: various schemes employing trapped ions, cavity quantum electrodynamics, nuclear magnetic resonance, neutral ions in an optical lattice, superconducting circuits, electrons floating on liquid helium, photon exchange interactions, and spintronics and quantum dots in solid-state systems. Although no solid state quantum computer implementation has yet been demonstrated, even for the simplest quantum gate execution, solid-state systems are thought to be most readily scalable to the required number of qubits, provided that the coherence time can be made sufficiently long. Quantum computer demonstrations based on magnetic resonance of J-coupled spin-$\frac{1}{2}$ nuclei within molecules in solution at room temperature have shown sufficiently long coherence times to be interesting. However, lack of an initial pure state at room temperature for these systems, because of the small nuclear Zeeman splitting, limits the scalability of this system. It has been proposed that similar systems employing electron spins isolated in solids at low temperature would have even longer coherence times while overcoming the pure initial state preparation issue. Although no demonstrations have been reported, this remains a promising approach to a useful and scalable quantum computer.

Artificial Brains with Natural Intelligence

Human brain function emerges from a complex network of more than a billion cooperating neurons whose activity is generated by nanoscale circuit elements. In other words, the brain is a massively parallel nanocomputer. For the first time, nanotechnology is revealing approaches to the design and construction of computational systems based more precisely upon the natural principles of biological nervous systems that include (1) enormous numbers of elementary nonlinear computational components, (2) extensive and interwoven networks of modifiable connectivity patterns, and (3) neurointeractive sensory and motor behavior.

A simple nanoelectronic component, the RTD, possesses the nonlinear characteristics that resemble the channel proteins responsible for much of the human neuron's complex behavior. At this time, such a direct nano-neuro analogy is most severely hampered by the truly difficult problem of interconnecting enormous networks of these nanocomponents. But as a beginning, this permits a consideration of the much greater density that is possible using nanoelectronic neurons than has so far been possible using microelectronic solutions, where equivalent chip architectures would need to be millions of times larger. Given their inherent nonlinearities, nanoelectronics may be better suited for such an artificial brain than for extension of today's digital applications.

Decades of neuroscience research have revealed the complexity of our brain's neuroarchitecture.[55] Despite this complexity, fundamental principles of organization have been established that permit a comprehensive sketch of our brain's functional neuroarchitecture. In addition, neuroscience has characterized many of the principles by which the network's connections are constantly changing and self-organizing throughout a lifetime of experience.[56] Indeed, the minute details of these trillions of connections cannot be specified uniquely, because they provide the basis whereby the unique qualities of each individual are structured by experience.

Until now, constrained by the limits of microtechnology, attempts to mimic human brain functions have dealt with the brain's extreme complexity by using mathematical simplifications (i.e., neural networks) or by carefully analyzing intelligent behavior (i.e., artificial intelligence). By opening doors to the design and construction of realistic brain-scale architectures, nanotechnology is allowing us to rethink approaches to humanlike brain function without eliminating the very complexity that makes this function possible in the first place. The tools of nonlinear dynamical mechanics provide the most suitable framework for describing and managing this extreme complexity.[57,58]

Instead, higher functions of the brain are emergent properties of its neuro-interactivity with the environment. Resorting to the notion of "emergence" always leaves an unsatisfying gap in any attempt to provide a complete explanation, but nature is full of examples, and classical descriptions of human nature have depended strongly on the concept of emergence (see Jean Piaget).[59] Specific functionality is not installed into the design of the system or programmed by a supervising algorithm, just as children, born with the natural intelligence for language, learn to communicate only by interacting with communicators. Generalizing, higher functions of a naturally intelligent complex system emerge as a result of mentored development within nurturing environments.

From this perspective, function is entirely self-organized and can only be interpreted with respect to the interactive behavior of the organism within meaningful contexts. For instance, speech communication develops by first listening to one's own speech sounds, learning to predict the sensory consequence of vocalization, and then extending those predictions to include the response of other speakers to one's own speech. At the core of this process is an active testing behavior that acts first and learns to predict the sensory consequences of each action within the context that generated the action.[60,61] Unfortunately, until we have a working model of such a natural intelligence, promises of humanlike intelligence remain speculative. But the potential flexibility and value of such an autonomous agent with a natural intelligence for communication, navigation, or exploration is worth the effort.

The possible applications of this new generation of humanlike machine intelligence are far-reaching, and it is difficult to imagine the ultimate impact. The obvious applications include (1) personal assistants and universal interpreters that communicate with natural language (e.g., speaking copilots interpreting complex situational data) and (2) autonomous human surrogates (e.g., unmanned vehicles that do not require telemetric control). In all such cases, an extended period of training would be necessary for the acquisition of such advanced humanlike abilities, but artificial brains might be copied, and since they are immortal, they might continue to improve. Also, optimization might employ evolutionary principles by copying the best into the rest and then reemploying the training to gain more improvements and so on.

Storage

The future of information storage is driven by the need for larger data volumes (and hence increased storage density), higher access speed, and lower cost. The eventual application of nanotechnology seems obvious because of the decreasing bit size of the stored information, the advantage of short interconnect distances, and the economy of parallel fabrication methods such as lithography being developed for computing hardware.

The quest for higher-capacity memories with faster access time will continue unabated for decades, as the capabilities of all operations associated with computers benefit. Cost, reliability, weight, and other characteristics continue to be important metrics. Besides the use of transistors in electronic circuits, material transformations such as melting or other phase change and magnetization are the most frequently used phenomena today. Current electronic memories (e.g., dynamic random-access memory (DRAM) and static random-access memory (SRAM)) use transistors. DRAM uses only one transistor and capacitor and requires refreshing many times a second to compensate for leakage, whereas SRAM uses more transistors (typically five) and does not require frequent refreshing. Both are fast, yet SRAM is considerably more expensive as a result of its larger size and number of components. Large amounts of data are stored on magnetic or optical disks, requiring moving parts that are subject to friction, wear, and the usual degradation associated with mechanical motion.

Magnetic Storage

Magnetic storage includes magnetic disks, tapes, and a new class of magnetic memory, random access nonvolatile memory (MRAM), with disks providing vast high-speed data storage, tapes providing archival storage of huge data sets, and MRAM providing a radiation-hardened, nonvolatile substitute for conventional electronic random access memory. Magnetic storage research and technology are related to nanotechnology in many ways. First, the increase in storage density has correspondingly reduced the magnetic bit size into the nanometer range. Perhaps more importantly, however, many of the length scales for magnetic interactions are characteristically on the nanoscale. For iron, for example, a domain wall width is 40 nanometers; an iron ferromagnet becomes a superparamagnet for particles less than 8 nanometers in radius at room temperature. In an iron-chromium-iron multilayer structure, exchange coupling reverses the relative magnetization of successive magnetic layers each time the chromium spacer layer thickness increases by 0.14 nanometers. Indeed, spin transport phenomena that occur on the nanoscale are responsible for the commercial success of giant magnetoresistive (GMR) magnetic read heads and magnetic field sensors. Historically, the aerial density of magnetic storage has increased exponentially at a rate roughly twice as fast as Moore's law for IC scaling. The aerial storage density, currently about 30 Gbit/in.^{2}, is expected to improve by a factor of 10 or more in the next 5 years.[62]

Research and technology development on nanotechnology applied to increasing the density of information storage on magnetic disks should be of importance in applications dealing with enormous data sets, such as may be obtained with multispectral imaging. Such work could involve improvements in GMR head sensitivity, the development of patterned magnetic media, finding ways to avoid or minimize the effect of bit sizes approaching the superparamagnetic size limit, and understanding tribology at the disk-head interface for nanoscale fly heights.

Magnetic random-access memories (MRAM) have been developed that typically employ a permanently magnetized pad, an electrically switchable magnetic pad to store the information, and a nanoscale, nonmagnetic interlayer. The value of the bit is read out by applying a voltage across the three-layer structure to determine the resistance. The resistance depends on the relative orientation of the two magnetic layers and is due to either the GMR effect, if the interlayer is a metal, or the tunneling magnetoresistance (TMR), if the interlayer is an insulator. The combination of a nonvolatile storage medium—that is, the magnetically aligned pad—and the ability to fabricate the entire structure by modest extensions of conventional semiconductor fabrication methods has led to the development of MRAM chips of modest storage capacity. Further development should lead to greater storage capacity and speed. Such memory devices would be useful in satellites, for example, or wherever nonvolatility, i.e., zero standby power consumption, is important. The technology is now being commercialized.

Physical Storage

Imagine having the ability to place atoms where you want them, one at a time. Think of the enormous storage capacity you would have if you could read and write atoms at specific sites separated by atomic scale distances on atomically flat surfaces. Such speculation notwithstanding, recent developments in nanotechnology point to the use of nanostructures, as opposed to single atoms, to store and read information. In one scheme, an atomic force microscope (AFM) is used to modify a surface as well as to sense the modification. Because each process would be relatively slow, a practical device would have to employ arrays of AFMs acting in parallel. One implementation of this idea, being pursued by IBM,[63] uses 1,024 cantilevers to impress 30- to 40-nanometer indentations in a PMMA layer, which corresponds to a storage capacity of 400-500 Gbit/in.^{2}

Molecular Memories

The smallest bit of matter that can retain a memory is an atom. Although it is possible to read and write information by placement of individual atoms, the time required to do so is prohibitively long. The possibility of using molecular bits of matter along with chemical transformations could be an attractive alternative. Electrical conductivity and optical density of a polymer or other material change significantly with chemical changes. Such changes may be induced photochemically, electrically, physically (pressure, temperature), or by other means of "switching," and these means have been examined extensively. With the increased interest in nanotechnology, the scaling of these effects to nanometer dimensions is being explored. Access speed, serial or random access, reliability, and volatility are current issues. While the fundamental properties of these alternative memory systems are being investigated, progress continues with conventional DRAM and SRAM approaches as miniaturization advances.

Single-molecule conductivity exhibits fascinating negative differential resistance current–voltage characteristics.[64] The origins of this behavior are not yet well understood. Negative differential resistance does form the basis for memory functions. Methods to access large numbers of single molecules (or even nanostructures), speed, retention times, etc. must be explored before molecular memory can become a viable hardware approach.

Another example of a molecular memory device is worth noting. A recent patent by researchers at Hewlett-Packard[65] makes use of hysteresis in the current–voltage curve observed for certain molecular materials sandwiched between two crossed conducting wires. This structure can be made arbitrarily (within the dimensions of a molecule) small. With such phenomena, a memory can be readily envisioned whereby the conductivity between two wires in a crossbar architecture can store and retrieve a bit of information. The retention time of this behavior is sufficiently long (many seconds, or even minutes for some materials) that such a device could have advantages over nonvolatile memories involving magnetic storage. The exact nature of the hysteresis observed is not yet understood. The ease of manufacture and the competitive nature of such a device compared with magnetic memory crossbar structures and with conventional magnetic disks have yet to be determined.

Yet another process using molecular memory is holographic information storage, where data are stored in a volume that retains optical phase information through a material transformation (typically an absorption changing the electronic state of a dye such as bacteriorhodopsin.[66] Although the information density for holographic storage is often quoted in Gbit/cm^2 for comparison with conventional storage, holography is inherently a volumetric technique with intrinsic units of Gbit/cm^3. Holographic storage is inherently analog and requires extensive electronics to retrieve digital information. Intensive research efforts to develop holographic information storage have yet to produce a commercial result.

Communications

Communications are a critical aspect of information technology. The Air Force's requirements are for very large aggregate bandwidths between large numbers of time-varying platforms with free-space optical, microwave, or RF connectivity. The overall system must be both dynamic and reconfigurable. Platforms will continuously move into and out of the system as a result of operations. The total amount of data will become staggering as large numbers of complex sensors (e.g., multispectral focal plane arrays) are fielded. Two consequences are immediately evident: (1) it will be necessary to provide a hierarchical structure to ensure that time- and command-critical messages are prioritized and (2) the sensors must be intelligent, with the capability of drawing inferences from raw data and transmitting condensed information and of switching to a more voluminous raw data stream as critical data fusion requirements demand. Discussion of these systems aspects is deferred to the next section, "Signal and Information Processing and Data Fusion." Here, the hardware aspects of communications are discussed.

Secure Communications

An emerging benefit of some of the physics of nanotechnology is communications security and secrecy. The application of certain quantum principles allows communications to be made perfectly secure in the sense of not allowing a message to be listened to without detection. One can prepare the quantum states of individual photons, e.g., polarization, in such a way that the "no cloning theorem" prohibits their being read without detection yet allows a "quantum repeater" to relay the signal to a distant source. Since such a scheme can be jammed, it does not seem attractive for general secure communications. However, it may have considerable value in periodically distributing the keys needed to decode encrypted messages sent by more conventional means. Of particular interest to the Air Force would be the use of such a secure form of communication to send encryption keys to satellites. Research on sources of single photons on demand and single photon detectors might be valuable.

Hardware aspects of communication will be enhanced by miniaturization trends in general simply through the introduction of denser and faster digital electronics. Advances in miniaturization will directly affect hardware design for military communication systems. Secure communications will take place through hardware and software designed for the purpose. Compressing the essential data in order to transmit with minimum bandwidth requires knowledge of the data being transmitted and is largely a software development.

Optical materials and devices are used in many communications operations. These include photonic switches and waveguides that may be fabricated with greater precision and smaller size using materials and processes emerging from research in nanotechnology, such as vertical cavity surface-emitting lasers (VCSELs), quantum dot lasers, and photonic bandgap materials.

New materials for both optical and RF applications are emerging from continuing investigations of semiconductors. Two examples are self-assembled quantum dots for laser systems and gallium nitride (GaN) and related materials for both optical (and ultraviolet) emission and high-power RF devices. These materials are discussed in Chapter 4.

Both optical and RF MEMS devices are of growing importance for communications and are discussed below.

Optical Devices

Micro- and nanotechnologies have already had significant impact on III-V, fiber, and nonlinear optoelectronic devices. For III-V devices, control of growth thickness using modern epitaxial growth techniques such as molecular beam epitaxy (MBE) and metal-organic chemical vapor deposition (MOCVD, also known as organometallic vapor-phase epitaxy, OMVPE) has progressed to the point that individual atomic layers (~0.5 nanometer thick) are routinely grown with differing material compositions. Figure 3-4 shows the evolution of the lowest threshold current for cw diode lasers.[67] It gives some perspective on the impact of improved growth capabilities and the advantages of nanostructures for optical devices. The first diode lasers, demonstrated in 1962,[68] were composed of homojunction materials (p-n junctions in a single material, GaAs). They were pulsed devices that were incapable of cw operation because of the power dissipation associated with inefficiencies inherent in this simple structure. The first cw lasers employed double heterostructures, where an index step was built in by varying material composition to confine the optical field to the junction region and dramatically lower the optical losses. The first data point on the double heterostructure curve is due to Alferov and was part of the work for which he was awarded the Nobel Prize in Physics in 2000. The scale, in the growth direction, of the double heterostructure waveguide is on the order of several hundred nanometers. The next major breakthrough was the addition of quantum wells to confine the carriers in the center of the optical mode and modify the electronic wave function and density of states as a consequence of the confinement. These quantum wells are on the scale of the de Broglie wavelength of the electrons and are typically tens of nanometers thick in the growth direction. The most recent curve

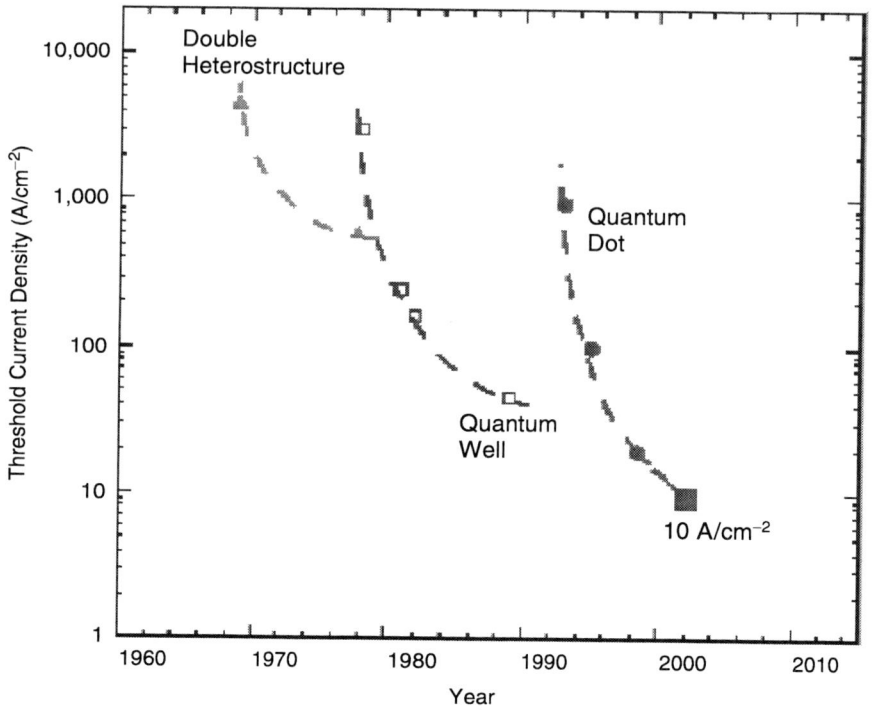

FIGURE 3-4 Diode laser thresholds. SOURCE: Adapted from Ledentsov, N.N., M. Grundmann, F. Heinrichsdorff, D. Bimberg, V.M. Ustinov, A.E. Zhukov, M.V. Maximov, Z.I. Alferov, and J.A. Lott. 2000. Quantum dot heterostructure lasers. IEEE Journal of Selected Topics in Quantum Electronics 6(3): 439–451. © 2000 IEEE.

corresponds to three-dimensional confinement of the electrons in quantum dots for which all three dimensions are on the nanometer scale. Quantum dots will be discussed in more detail below; for the nonce it is sufficient to appreciate that nanoscale confinement in three dimensions serves to further modify the electronic wavefunctions and density of states, now to an atomic-like discreteness, and to increase the local electronic density of states.

As is clear from this brief overview, micro- and nanotechnologies are already having dramatic effects on optoelectronics. Modern telecommunications is based on quantum-well lasers, optical fibers and waveguides, modulators, detectors, wavelength- and time-division multiplexers, etc.—all of which are based on structures that provide for careful control of electronic and optical wavefunctions, and all of which are on the nanometer scale.

Electronic Confinement. Quantum wells have been investigated since the early development of epitaxial growth capabilities. Nonetheless, new innovations continue to show that we have not yet fully exploited this capability. There is, for example, active research on mid-infrared diode lasers using both Sb-based III-Vs mixing type I (both electrons and holes confined) and type II (electrons confined, holes not) band alignments,[69] and intraband quantum cascade configurations.[70] These sources have direct application to Air Force countermeasure needs as well as (potentially) to free-space optical communications (free-space transmission distances are much longer in the 8-12 atmospheric window than in conventional 1.3- and 1.55-micrometer telecommunications bands) and to both commercial and military needs for remote sensing of gases.

Digital alloy growths composed of multiple layers of alternating material composition, typically ≤10 nanometer thick, provide another planar (growth direction) nanostructuring approach that is increasing in importance. Using digital alloys it is possible to create materials that mimic quaternary or even more complex III-V materials while only growing binary and ternary compounds where the growth control is much improved. It is even possible to synthesize material layers that are inaccessible to conventional multielement growth because of growth kinetic limitations such as miscibility gaps. These layers are certain to find applications in both optical and electronic devices.

As noted above, two-dimensional (quantum wire) and three-dimensional (quantum dot, or QD) confinement, requiring small dimensions in the growth plane as well as in the growth direction, are exciting areas of very active research. For the most part, the fabrication of these materials is based on self-assembly, specifically the Stransky Krastanov growth[71] mode, where the interplay between lattice mismatch stress and surface tension of the growing film leads to the formation of isolated islands or quantum dots. Figure 3-5 shows an atomic-force microscope image of an array of InAs quantum dashes on InP with a photoluminescence peak at 1.6 mm, matching the long-haul telecommunications band.[72] As in all electronic devices, the material demands are very exacting. In particular for quantum dots, the interface between the quantum dot material (often InAs) and the surrounding semiconductor (InGaAs or GaAs) has a strong impact on the optical and electronic properties. To date, attempts to use top-down lithographic patterning have resulted in defects leading to nonradiative recombination and an optically "dead" device. However, the advantages of a monodisperse QD size distribution along with a well-defined positional arrangement would be dramatic, and many different approaches are being pursued. Achievement of this goal would result in significant improvement in many optical devices, specifically, a dramatic reduction in the threshold current density and vastly narrowed spectrum. Spatially tuning the dot size would give a distribution of sources for wavelength division multiplex applications. Quantum dots are also being explored as qubits for quantum computing applications; they would require a coupled dot system with the ability to electronically tune the coupling. This is far from today's reality, but not unreasonable to envision given the continuing advances in both lithography and crystal growth discussed elsewhere in this report. Quantum dot

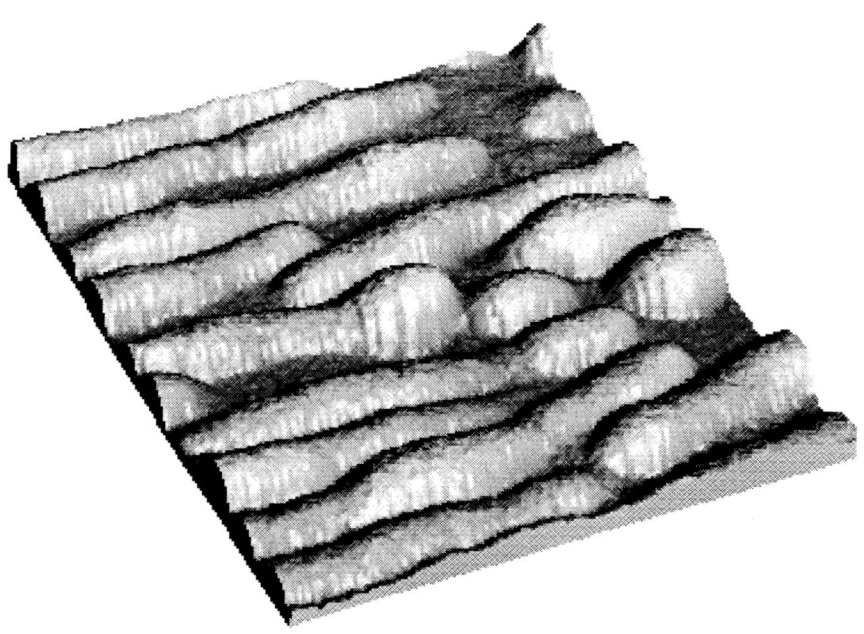

FIGURE 3-5 InAs quantum dashes grown on InP. SOURCE: Wang, R.H., A. Stintz, P.M. Varangis, T.C. Newell, H. Li, K.J. Malloy, and L.F. Lester. 2001. Room-temperature operation of InAs quantum-dash lasers on InP. IEEE Photonics Technology Letters 13(8): 767–769. © 2001 IEEE.

lasers have exhibited relatively high power; the inherently low linewidth enhancement factor, or alpha-parameter, suggests that they may be suitable for scaling to much higher powers than conventional diode lasers while still retaining coherence (e.g., improved brightness and power).

The science and technology of three-dimensional nanostructured semiconductors is just at its beginnings. Looking backwards, improvements in control of semiconductor materials always have led to improvements in electronic and optical devices and to new and improved systems applications of those improvements. There is every reason to expect that today's developments will lead to tomorrow's devices—with vastly improved characteristics that will enable new applications for both military and commercial markets.

Optical Confinement. The characteristic length for optical confinement is $\sim\lambda/2\sqrt{\Delta\varepsilon}$, where λ is the optical wavelength and $\Delta\varepsilon$ is the index difference between the guiding and cladding regions. For semiconductor/air interfaces, this reduces to $\sim\lambda/2n$ (n is the semiconductor refractive index) well into the nanometer regime for typical values ($\lambda \sim 1$ micrometer, $n \sim 4$). The history of optical confinement structures parallels that of electronic confinement. The earliest structures were planar waveguides (one-dimensional) and channel waveguides (two-dimensional) where the second dimension was well in the micrometer regime owing to the use of low index contrast in the plane and accessible by conventional lithography. Optical fibers are a ubiquitous example of two-dimensional waveguides that have had a major impact on communications and sensing technologies.

One-dimensional Bragg gratings (layered structures where the reflection is built up by the in-phase addition of many small amplitude reflections) have long been used for frequency control of semiconductor lasers. Distributed feedback and distributed Bragg reflector lasers are two examples of the use of Bragg gratings.[73] The discovery of photosensitive writing of Bragg gratings directly into telecommunications fibers has found many uses,[74] especially with the rapid development of wavelength division multiplexing (WDM) technologies to increase the information-carrying capability of telecommunications fibers. Another use of Bragg reflectors arose with the development of vertical-cavity surface-emitting lasers (VCSELs).[75] VCSELs have many advantages over conventional edge-emitting lasers and are rapidly becoming a dominant technology for data communications applications.

An emerging application related to Bragg gratings is periodic poling of materials (to date mainly $LiNbNO_3$ and related ferroelectrics).[76] The motivation in this case is the requirement for phase matching in nonlinear optical processes, the goal is a variation in the nonlinear optical properties (e.g., the ferroelectric polarization) rather than in the linear (index) response, and the scale of the structure is set by the coherence length between the various optical waves (typically this scale is larger than the wavelength and easier to fabricate with conventional methods). Periodically poled materials are now being investigated for systems applications such as optical mixers, analogous to the RF mixers that have found extensive application in communications.

Research is now moving to higher dimensionality structures known as photonic crystals as a result of the strong analogy between the behavior of photons in a periodic dielectric lattice and electrons in a periodic potential lattice.[77] There are both fully three-dimensional versions of photonic lattices[78] and two-dimensional versions where the in-plane confinement is provided by a two-dimensional dielectric array (often, holes in a semiconductor) and the out-of-plane confinement by a conventional dielectric discontinuity (waveguide slab). The two-dimensional variety is easier to fabricate, and so it is attracting much attention. As fabrication skills improve, there will be increasing interest in three-dimensional photonic crystals. By building nanocircuit elements (e.g., small tank circuits with both capacitive and inductive components), it has recently been predicted[79] and demonstrated[80] (with millimeter waves) that truly novel materials are possible—for example, materials that exhibit negative refraction, in which light focuses rather than diverges and for which diffractionless imaging is theoretically possible. With history as a guide, there is little doubt that the development of photonic crystal materials will have a major impact on optical applications.

Microelectromechanical Systems (MEMS)

MEMS are structures, devices, or systems having some parts on the scale of micrometers that are produced by any of several techniques collectively termed "micromachining." They are generally called microsystems in Europe and, sometimes, micromechatronics in Japan. In broad terms, there are four classes of MEMS. The first involves micromachined structures that have no moving parts. Included here are channels and nozzles for fluids and guides for optical and RF signals. The second class of MEMS is sensors that transduce some aspect of the world into electronic data. Many MEMS sensors have been commercialized, notably pressure sensors, microphones, accelerometers, and angular rate sensors. The third class of MEMS includes mostly actuators, essentially the inverse of sensors, because they transduce information into some physical, chemical, or biological effect. They are now being made into products, primarily for communications. The last class of MEMS includes systems that involve both sensors and actuators. Microfluidic systems already on the market, data storage systems that promise terabit per square centimeter densities, and micro-energy systems now under development are in this class.

Optical MEMS. Optical MEMS, sometimes called MOEMS, involve light in the visible, infrared, or ultraviolet spectral regions. There is a natural aspect to the interaction of light with MEMS, because the dimensions of MEMS, even though small by ordinary standards, are large compared to the wavelength of optical radiation. Hence, micro-optical systems can manipulate light in a diffraction-limit regime. Further, light exerts negligible pressure on MEMS components, so that only small forces are required for moving and holding mirrors and other optical structures. Optical MEMS are poised to have a dramatic impact on the optical networks that are the backbone of the Internet. In fact, they could be an enabling element for the "all-optical" network that would eliminate the need to convert from optical to electronic and back to optical at switching nodes. Micromirrors will reflect signals from an incoming optical fiber, whatever their wavelength, to the proper outgoing fiber. There is still electronics involved in reading the headers on the incoming optical signals and powering the MEMS devices; this network is not based on all-optical nonlinearities, which are generally too weak for practical use. Other approaches to optical switching include microfluidics and attenuated total internal reflection, both of which are alternative MEMS techniques, and electrooptic switches based on either inorganic (e.g., $LiNbO_3$) or polymer organic electro-optic materials. Given the wide range of requirements and applications—from high-speed modulation to individual wavelength provisioning and long-haul to local-area networks—it is likely roles will be found for multiple technologies.

A convenient way to classify optical MEMS is according to their involvement in the birth or life or death of photons—that is, as sources, optical transmission and switching, or detectors. Figure 3-6 shows examples of MEMS from each of these classes, including two types of MEMS mirrors. The adjustable end mirror of a vertical-emitting-cavity, solid-state laser source is on the upper left. In the upper-right image, the needle is resting on an array of mirrors for fiberoptic signal routing made by Lucent Technologies. One of the mirrors is shown in a

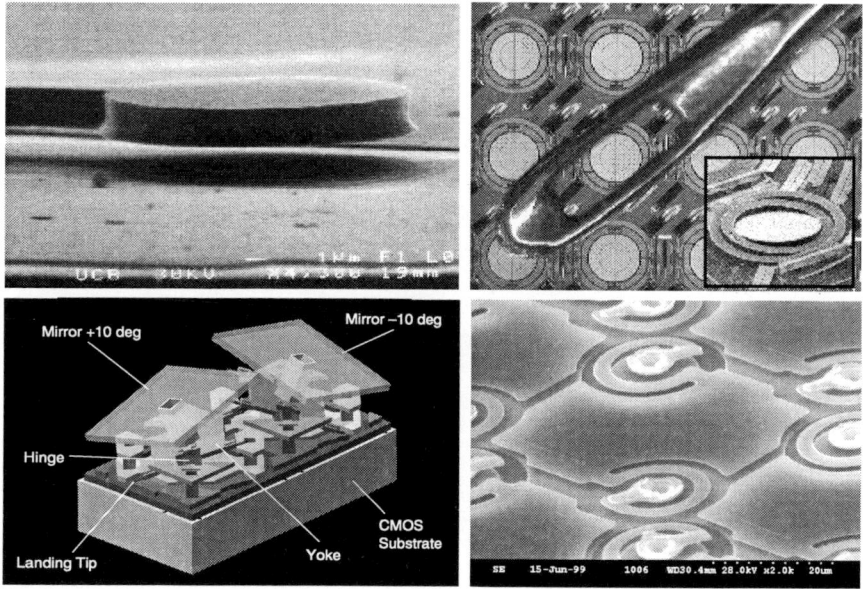

FIGURE 3-6 Optical MEMS examples. SOURCE: *Top left:* Chang-Hasnain, C., E. Vail, and M. Wu. 1996. Widely tunable micromechanical vertical cavity lasers and detectors. Proceedings of the IEEE/LEOS Summer Topical Meetings on Optical MEMS and Their Applications. New York, N.Y.: Institute of Electrical and Electronics Engineers, Inc. © 1996 IEEE. *Top right:* Bishop, D.J., C.R. Giles, and S.R. Das. 2001. The Rise of Optical Switching. Scientific American 284(1): 88–95; image courtesy of Lucent Technologies' Bell Labs. *Bottom left:* Available online at <http://www.dlp.com/about_dlp/about_dlp_images_dlpcinema.asp> [July 30, 2002]; *Bottom right:* Schimert, T.R., J.F. Brady, S.J. Ropson, R.W. Gooch, B. Ritchey, P. McCardel, A.J. Syllaios, J. Tregilgas, K. Rachels, M. Weinstein, and J. Wynn. 2000. Low power uncooled 120 × 160 a-Si-based micro infrared camera for unattended ground sensor applications. Pp. 23–30 in Unattended Ground Sensor Technologies and Applications II, Proceedings of SPIE, Vol. 4040, E.M. Carapezza and T.M. Mintz, eds. Bellingham, Wash.: The International Society for Optical Engineering; image courtesy of Raytheon.

tilted position in the inset. The drawing on the left shows two of the 16-μm² pixels from the digital mirror device sold by Texas Instruments (more detail on the fabrication of this device is given in Chapter 4). A micrograph of part of an array of thermoresistive pixels from a Honeywell uncooled infrared imager is also shown (lower right).

The source is a VCSEL in which the cavity length, and hence the output wavelength, can be tuned by moving a micromirror on a cantilever with a voltage-determined electrostatic force.[81] VCSELs are becoming the source of choice for optical fiber networks, and the tunability of such devices is useful for wavelength agility.

Micromirrors have been made in many different configurations individually and in arrays. The round mirrors in Figure 3-6 were developed by Lucent Technologies for switching signals from one optical fiber to another.[82] They are 500 micrometers in diameter and can be tilted controllably in two directions. Such mirrors have been made into arrays for switching optical pulses between 1,024 incoming fibers to the same number of exit fibers. The system containing the MEMS mirror array, termed a lambda router, is capable of passing all the current Internet and telephone traffic in the world.

Alternative MEMS optical switches based on switchable gratings and on fluidic devices (bubbles) are also under development. Small as well as large companies are involved in developing MEMS for network switching. In 2000, four start-up companies making optical MEMS mirrors were sold for a total of almost $6 billion in stock; in the spring of 2002, their value was much less. The decline in the stock market and in the pace of telecommunications infrastructure development has delayed the introduction of MEMS switches into the long-haul communications grid, but they are now beginning to be used. MEMS mirror arrays are proving important not only for the communication of information but also for its display. A digital mirror device (DMD)[83] is the display engine for conference room projectors and, soon, for digitally fed movie theaters and home entertainment centers. The micromirrors in the DMD flip ±10 degrees to pass the incident light to the screen or into a beam dump. The device consists of three layers, the lower being electronic, the middle having the mechanics, and the top being the optical mirror. The torsional hinges supporting the mirrors are 5 × 1 × 0.1 micrometers, and they last for over 10 billion cycles. Each of the mirrors is 16 mm on a side. DMDs now on the market contain millions of mirrors. Of all the systems ever produced by humankind, these devices have the most moving parts. It has been reported that 100,000 of these devices have been sold in the past few years, so there are over 50 billion MEMS mirrors already in use for this single optical MEMS display technology (see Chapter 4 for a discussion of the manufacturing challenges that had to be surmounted).

The last type of optical MEMS is detectors, which are generally found in arrays to form imagers. They are particularly important for the imaging of objects in the infrared (IR) region, where the heat from people and ordinary objects is especially useful. IR detector arrays are made by Raytheon and other companies. They cost about one tenth of the cost of the military IR detector arrays based on narrow bandgap semiconductors, which require cooling to low temperatures. The uncooled MEMS IR arrays are expected to find widespread civilian uses—for example, in automobiles to provide images of people and animals in the road beyond the range of normal headlights.

Optical MEMS, especially those for the transmission of information and display of images, have come on the scene as the Internet has become an indispensable part of modern life for individuals, schools, businesses, organizations, and governments. In a similar fashion, RF MEMS are emerging from laboratories into commercial use right in the middle of the wireless revolution. Soon, one in six people on Earth will have a cell phone (in a world where half the people have never made a phone call!). It is projected that one in twenty people will soon have wireless internet connectivity.[84]

Radio frequency MEMS. Many types of RF MEMS components have been demonstrated in recent years. They include waveguides, detectors, and inductors, which do not have moving parts, and capacitors, switches, and resonators, which are micromechanical devices. There are many kinds of micromechanical RF MEMS capacitors; two are shown in Figure 3-7. The three-dimensional interdigitated structure on the left shows a capacitor developed by the Rockwell Science Center for tuning an RF circuit, made by bulk micromachining.[85] The "bowtie"

 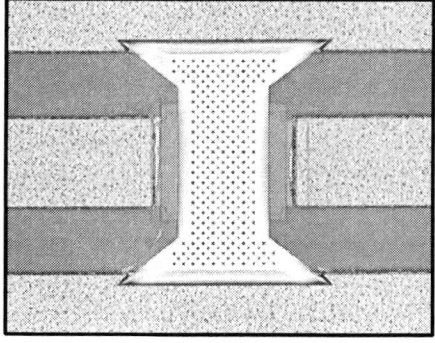

FIGURE 3-7 RF MEMS capacitors. SOURCES: *Left:* Yao, J.J., S. Park, and J. DeNatale. 1998. High tuning-ratio MEMS-based tunable capacitors for RF communications applications. Technical Digest of the Sensors and Actuators Workshop. Cleveland Heights, Ohio: The Transducer Research Foundation, Inc. *Right:* Yao, Z.J., S. Chen, S. Eshelman, D. Denniston, and C. Goldsmith. 1999. Micromachined low-loss microwave switches. IEEE Journal of Microelectromechanical Systems 8(2): 129–134. © 1999 IEEE.

device on the right, shown in a top view, made by Raytheon to shunt RF signals from the center conductor to the upper and lower grounds when it is pulled down, was produced by surface micromachining.[86] The center strip is about 50 mm wide. Micromechanical capacitors perform two basic functions. The first is simply to provide voltage-controllable capacitance changes in an RF circuit. The second is to switch RF signals in tunable filters and other subsystems. The devices in Figure 3-7 serve these two functions.

RF signals can also be switched by MEMS devices that provide metal-to-metal contact. Several such switches have also been prototyped. These devices involve challenging materials problems. The reliability of RF MEMS switches is an issue key to their commercial utility. Raytheon has operated the capacitive switch shown in Figure 3-7 for over 100 million cycles.[87] Motorola has run an RF MEMS switch with metal-to-metal contacts for 4 billion cycles.[88] The most demanding applications of micromechanical switches for RF and microwave applications will require performance to 100 billion cycles.

Microresonators are another promising type of RF MEMS. They provide frequency standards in transmitters and receivers, similar to the quartz crystals now used in RF and microwave systems. The great promise of micromachined resonators is the possibility of integrating them on the same chip with CMOS electronics, thus reducing the parts count and the shock susceptibility of systems. Such integration is one of the main challenges facing the utilization of these resonators. They must also be vacuum packaged without any atmosphere, which would degrade their performance. Such packaging must protect the resonators for as much as a decade in use. Cleanliness within the package to defeat desorption of molecules onto the microresonators, which would cause frequency shifts, is another requirement.

MEMS pressure and acceleration sensors were sold by the tens of millions in the 1990s. It is likely that MEMS actuators for optical and RF applications will enjoy similar success in the current decade. Optical MEMS will find greatest use for the fiber-optic communication of information and for the display of images.[89] RF MEMS will be widely used in cell phones and other wireless communications equipment.[90]

Signal and Information Processing and Data Fusion

Without intelligent processing of data, the Air Force will not be able to take advantage of the advances in sensing and communications. Indeed, there is danger of being buried in data, of all of the communications pipes being fully engaged in transmitting reams of bits—mostly of low value—and of a severely restricted ability to respond.

The functions of communication and data fusion represent aspects of information technology that are user unique. These functions demand considerable attention by the Air Force simply because of the highly specialized nature of its mission. The development of more capable digital systems will be driven by their widespread commercial applications, while specialized hardware and software for communication and data fusion will be highly dependent on the specific nature of the sensors and the missions to be performed.

It is clear that miniaturization will proceed for at least several orders of magnitude in computing power with continuing advances in scaled CMOS. As the other electronic systems being investigated come to fruition, this advance is likely to continue and—perhaps—even accelerate. Using all of this power to sense, to communicate, and to compute to maximum advantage is a systems challenge of the highest order.

Figure 3-8 is a schematic representation of the components of a situational awareness system. All of the computation requirements (at the sensors, at the distributed command and control nodes, and at the response platforms) will need to be exceptionally robust and resistant to enemy attempts at disrupting information. The communications links will have to be exceptionally robust. Large amounts of data will have to be transmitted securely and reliably over a complex

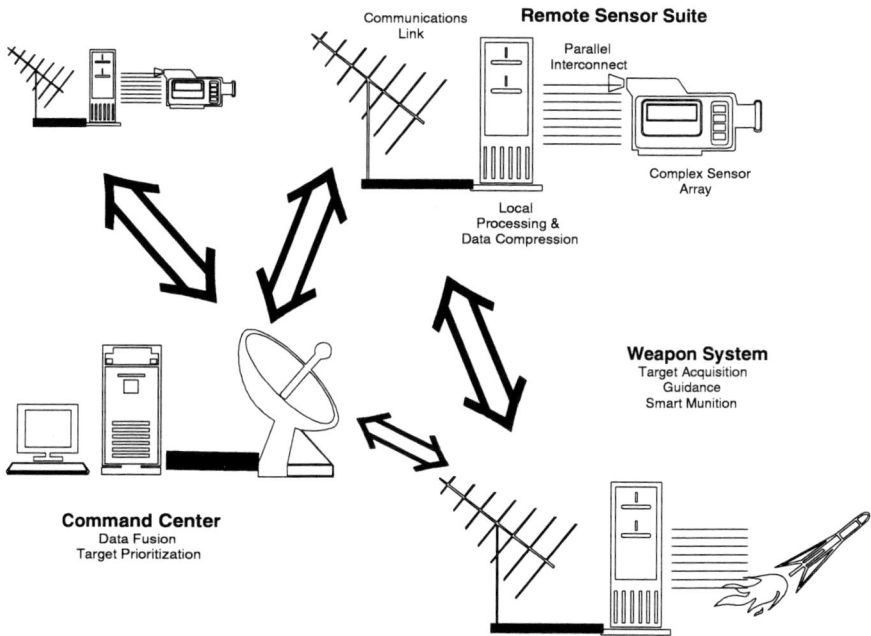

FIGURE 3-8 Schematic of a situational awareness system. SOURCE: National Research Council. 1999. Reducing the Logistics Burden for The Army After Next: Doing More with Less. Washington, D.C.: National Academy Press.

and rapidly changing network in the face of enemy attempts at dis-information and disruption.

The communication network will include many sources of very large amounts of data, many nodes trying to access information, many redundant communications channels (e.g., radio frequency, microwave, optical, earthbound, UAV, manned flight, and satellite platforms) to ensure connectivity, and many protocols (e.g., encryption, spread spectrum, frequency hopping, code division multiplex, and ultrabroadband) to ensure authenticity and increase aggregate bandwidth. Network management in such a complex environment will require robust software to ensure that the system can operate reliably under highly stressful conditions. Commercial systems are based on the very democratic Internet protocol (i.e., the system slows down as more users demand access). A military system must have a more hierarchical structure based on dynamic prioritization. The commander needs a continuous overview. A sensor that detects an imminent threat must be able to get its message across. This is a complex networking problem that will be difficult to solve.

The committee was not presented with any information on activities in software and systems along these lines during several meetings with AFRL and Air Force Office of Scientific Research (AFOSR) personnel. The Air Force must develop a strong capability in software and embedded systems integrating software and hardware if it is to be able to capitalize on the dramatic advances in hardware discussed elsewhere in this report.

Data Fusion

Data fusion is largely a software development, although it also requires display devices that maximize the transmission of key concepts and ideas to a decision maker. The ability to gather data from multiple sensors and to manipulate the data to get the best information will increase along with the capabilities of computers.

To reduce the communication demands of increasingly capable sensors, local intelligence will be required at the sensor to preprocess the data and reduce the bandwidth requirements for transmission. A single multispectral (5 bands), real-time (60 frames/sec), high-resolution (10 megapixels; 10 bits/pixel) video signal requires as much as 30 Gbit/sec bandwidth before compression. Multiply this demand by perhaps thousands of such sensors and the problem becomes clear. The volume of data to be processed will require sophisticated computation for automated analysis and semiautomated or even autonomous decision making and rapid response. There must be a dynamic allocation of the information burden between local and centralized computing and communications. The optimal admixture is clearly situation dependent and must be dynamically allocated by the software. After target acquisition, decisions must be communicated to the appropriate weapons platform. Sensor feedback on battle damage must be assessed and intelligence data updated.

Software and Codesign

Software and integration with hardware are serious and growing problems. There are significant issues related to software complexity, stability, reliability, security, and adaptability. The Air Force will require a very complex software overlay for assimilating terabits of data from a large number of sensors in a rapidly changing environment and intelligently distilling the resulting information for users as varied as commanders and autonomous platforms. Fundamental questions must be answered about the limits of robustness and stability in such a system.

Future embedded systems will be software-dominated. "Our technological environment will become, step by step, more and more networked, more and more autonomous, and more and more self-organizing."[91] The industry perspective on the migration of software technologies is illustrated in Figure 3-9.

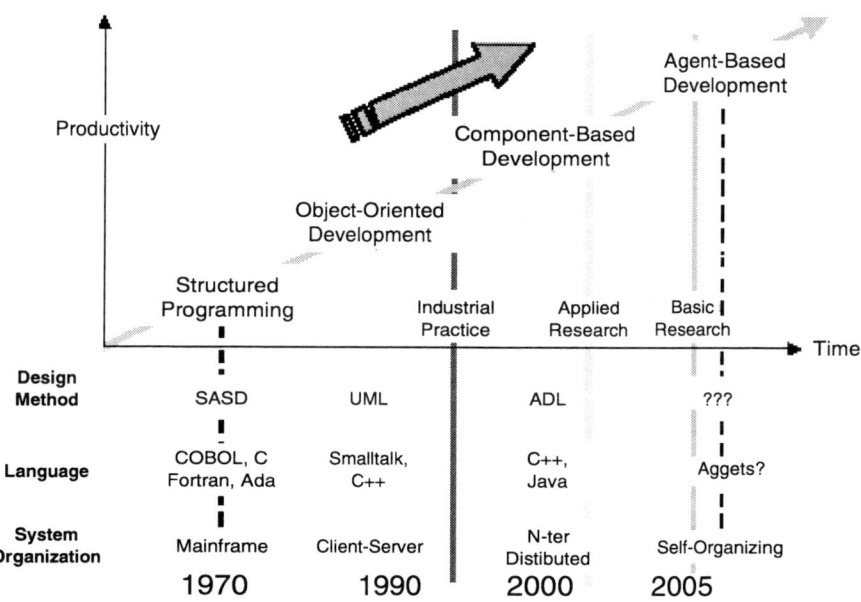

FIGURE 3-9 Paradigm shifts in software. SOURCE: ITEA Office Association. 2001. Technology Roadmap on Software Intensive Systems: The Vision of ITEA (SOFTEC Project), March. Available online at <http://www.itea-office.org> [July 31, 2002]. © 2001, ITEA Office Association. All rights reserved.

The engineering process will require new methodologies, models, and implementation techniques to support the efficient implementation of end-to-end services. Engineering technology will require techniques to master complexity and capture end-to-end specification of a distributed system. Implementation tools that guarantee a cost efficient, resource-limited implementation on a wide variety of platforms are required. A design flow that supports the development process in a multidisciplinary environment will be necessary. The move to component-based systems (e.g., to increased software reuse) will lead to new opportunities and to new demands on the Air Force software system.

Findings and Recommendations

The findings and recommendations considered by the committee to be critical appear in the Executive Summary. They are designated T (for technical) or P (for policy). More specific findings and recommendations, which are not included in the Executive Summary, are numbered to reflect their order within the report's chapters. All the findings and recommendations are collected in Chapter 7.

Finding T1. *Further miniaturization of digital electronics with increased density (~128×) is projected by the integrated circuit industry over the next 15 years based on continued scaling of current technology.* The most recent ITRS forecasts the accelerated introduction of smaller dimensions and greater computational power than were forecast by the Information Technology Roadmap for Semiconductors of 2 years ago.

Recommendation T1. *The Air Force should position itself to take advantage of the advances predicted by the Information Technology Roadmap for Semiconductors.* Dramatic advances are predicted for device technology. Software, application-specific integrated circuits (ASICs), embedded computers integrating software and hardware for specialized applications, and radiation-hardening and packaging for hostile environments must be designed by, and for, the military, to take advantage of these advances.

Finding T2. *In anticipation of an ultimate end to the historical scaling of today's integrated circuit technology, many new and alternative concepts involving nanometer-dimensioned structures are being examined.* As yet, none of these concepts had demonstrated the necessary functionality and integrability to be a clear choice for "beyond silicon." Many different material and device technologies will need to be explored well into the future. Two facts seem clear. First, it is not possible to make reliable, long-term predictions of breakthrough capabilities emerging from the rich frontier of discovery, fabrication, and material properties at nanometer dimensions. The numerous avenues of research investigation are likely to uncover unexpected processes and/or material properties that will have an impact at the fundamental level of information processing. Second, it seems likely that the initial applications of any of these technologies will build on and enhance the very strong base of existing integrated circuit technology, which will provide the necessary backbone of functionality and integrability until an entirely new computation paradigm emerges.

Recommendation T2. *Exploration of the scientific frontiers involving new procedures for fabrication at nanodimensions and new nanoscale materials, properties, and phenomena should be supported. The Air Force should track, assimilate, and exploit the basic ideas emerging from the research community and continue to support both intra- and extramural activities.* The focus should be on understanding the fundamental processes for fabrication, and on the unique properties of materials and devices structures at nanometer dimensions. Extremely dense arrays of devices capable of manipulating bits rapidly and reliably should be a dominant aspect of these investigations. Individual devices with nanometer or molecular dimensions are demonstrating logical functions on a small scale with a limited number of examples. Molecular electronics appears promising at present. There is potential for new device innovations and for progress in computing architectures and strategies. Quantum computing and quantum cryptography are examples of the applications that may be enabled by further progress in micro- and nanotechnology. The technology may develop rapidly once the scientific principles and technological advantages are discovered and understood.

Finding 3-1. *Space electronics is vital to the Air Force mission. The unique characteristics of the exoatmospheric environment place special demands on electronics that are outside the mainstream developments of the integrated circuit industry.*

Recommendation 3-1. *The Air Force must maintain a research and development effort in radiation-hardened electronics and must evaluate the continuing developments of micro- and nanotechnology for their applicability to space.* Some commercial developments, such as the move to silicon-on-insulator (SOI) materials, have clear radiation hardness benefits; others, such as molecular electronics, have yet to be evaluated in this context but are likely to exacerbate the problems.

Finding 3-2. *Communication is a critical aspect of information superiority.* The continuing trend to miniaturization is evident in this area as well as in computation. Traditionally, communication has been dependent on a wider materials base than computation as a result of the need for optical interactions and for high-speed analog functions at microwave and RF frequencies.

The Air Force has unique communications requirements, different from those of the commercial sector, demanding sustained effort and the overlay of scientific advances with Air Force requirements.

Recommendation 3-2. *The Air Force should maintain a strong research program in both the optical and microwave/RF regimes.* MEMS technology is having a strong impact. Nanotechnology is already leading to advances in these areas as well as in computation.

Finding T6. *The Air Force strategic nanotechnology R&D plan, as presented to the committee, is focused on hardware concepts without appropriate consideration of total systems solutions.* It is well known that over the past 15 years the commercial sector has made increasing investments in architecture and software concepts to design advanced systems. The tendency has been toward codesign of the hardware and software aspects of a system. One implication of nanotechnology is that this approach will be even more essential as device capabilities continue to expand. New algorithms, architectures, and software design methods will need to be developed and employed in concert with new nanotechnology-based hardware. Investment in this strategy will enable autonomous, intelligent, self-configuring Air Force systems. The Air Force strategic plan contains many future scenarios where such systems would be the ideal, if not the only, solution.

Recommendation T6. *The Air Force should take seriously the importance of co-system design as a critical consequence of continued miniaturization and should invest in the algorithm, architecture, and software R&D that will enable the codesign of hardware and software systems. This should be undertaken along with a projection of the advances that will be made in hardware.*

SENSORS

Introduction

Sensors are the eyes, ears, and nose of military systems, acquiring and processing data to enable decision making and actions by war fighters. Information from sensors is collected from U.S. Air Force and other sources, analyzed, compiled, and disseminated to other military branches and to our allies. Militarily useful categories of sensors are familiar: They include electromagnetic spectrum sensors and imagers, acoustic and seismic sensors, inertial and position sensors, and sensors of specific chemicals and compounds. The effectiveness of a military information system depends crucially on the intelligent integration of sensor systems with the signal processing, computing, and communications systems discussed in the previous section.

The current trend toward integrating data into a total battlespace "infosphere" consisting of a common operating picture, interoperability of systems, seamless integration of space, air, and ground assets, and knowledge management leading to decision superiority will in many respects require further developments in sensor systems. To achieve this necessary state, we need to have robust, sensitive, high-resolution, broadband detection, as well as appropriate preprocessing and instantaneous secure transmission of the important results for further processing. Nearly instantaneous decision-making capabilities and secure dispersal of the results of this process to the correct recipients no matter what platform or nation are also mandatory. And we need to maintain this overall capability all of the time, 24/7. Sensing clearly is key to our warfighting capability, before, during, and after the battle.

Discrete Versus Distributed Sensors

The U.S. Air Force desires "an AFRL NANO S&T Program providing revolutionary Air Force war fighting capabilities and preempting technology surprise."[92] Most sensors are designed to obtain narrowly defined information; examples include strain gages, thermometers, and accelerometers. These discrete sensors, when appropriately chosen, installed, and interrogated, provide a wealth of useful information. Recent developments allow different kinds of data taken from the same location to be combined into a sum of information greater than its constituent data. This is sufficiently important that the term "data fusion" has been coined to describe this combining of data from multiple sensors into one cohesive package. When this notion of multiple sensors is taken to extremes spatially, i.e., hundreds to thousands of sensors dispersed throughout an area of interest, the picture becomes more complete and nothing of interest goes undetected. One important direction for distributed sensor research is "smart dust,"[93] where large numbers of millimeter-scale, MEMS-based sensors communicating by wireless RF or optical links are dispersed over an area of interest. Clearly, networks such as these call for research into the size, functionality, and power consumption of individual nodes and into the communications and network architectures and hierarchy necessary to optimize their usefulness, especially in uncooperative environments.

Projected Impact

Near-term advances in sensor systems will lower the cost of the overall sensor systems, lighten the packages we fly, and eventually make some sensor types ubiquitous. Many of these near-term gains will come from the MEMS community as it improves its designs, fabrication processes, and materials.[94] In the medium term, we can expect that higher performance will be achieved from advances in nanomaterials and from increases in computing capacity, data storage, and secure communications, much of which will also be derived from the nanosciences. MEMS will allow us to produce large numbers of sensors more cheaply, but nanoscience will enhance the performance of these relatively inexpensive sensors.

A bit further out, perhaps by about 2020, we should see the successful mating of these improved performance sensors into systems of sensors and a resulting dramatic increase in useful understanding of the battlespace. Much of this expansion of sensor information will have been apparent earlier, but by 2020, we will begin taking this volume of information for granted.

Impact of MEMS

Within the scope of semiconductor-based devices, but outside the range of nanotechnologies and squarely in the range of microtechnologies, are MEMS-based sensors, including MEMS inertial sensors, magnetometers, room-temperature IR focal-plane arrays, acoustic and seismic sensors, and chemical and biological threat sensors. Of these, the MEMS pressure sensors and inertial sensors (accelerometers, angular rate sensors, and combinations of these) are by far the most mature. Pressure and inertial sensors were perhaps the major driving force for MEMS development at its earliest stages.

There is great motivation within the military for acoustic, seismic, magnetic, chemical, and inertial sensing capability in very small packages. The driving applications for this include unmanned air vehicles (UAVs), man-portable systems, compact and lower cost missiles, smart munitions, systems for detecting underground and hardened targets, smaller and lighter satellites and other spacecraft, and more highly integrated and generally capable avionics. In summary, all the pervasive motivations for smaller, more compact, more cost-effective, but still highly capable defense systems lead to MEMS sensors.

Impact of Nanotechnology

Many types of sensors are made using semiconductor materials, especially compound semiconductors. Thus, these sensors are a subcategory of microelectronics and optoelectronics. The impact of nanotechnologies on semiconductor-based sensors will be much like its impact on semiconductor-based information technologies. That is, nanometer-scale devices will offer significantly better performance in smaller and less expensive units.

The impacts of nanoscience on sensor technologies include improved materials, more precise monolayer-scale control of growth, novel heterostructures, and new device concepts such as quantum wires and dots or other quantum-effect designs. These are potentially at least as important for sensors as for the rest of the technologies based on micro- and nanoelectronics.

Nanometer control of epitaxial compound semiconductor multilayers used in complex semiconductor device structures is an area of technology that has made extraordinary advances in the last 20 years. In addition, in recent years, the development of heterogeneous integration techniques that combine multiple semiconductor families has greatly expanded the versatility and adaptability of less mature semiconductor materials and devices.

The heterogeneous integration of semiconductors, based on nanoscale control of fabrication processes, is beginning to impact sensors. One example is the self-assembly of large numbers of IR detector pixels into an array format on a curved substrate. Such schemes offer the potential for focal-plane arrays with extremely large fields of view and high resolution, which are required for effective, space-based operation.

Payload Sensors

Many sensing systems are an integral part of the mission payload. A large fraction of these payload sensors are imaging systems, where electric components will benefit—in reduced weight and volume and increased functionality—from developments in micro- and nanotechnologies as discussed in the previous section.

Electromagnetic Spectrum Sensors

The Air Force's role in space requires space-based smart sensors for battlespace surveillance,[95] including imagery and various modes of sensing. The goal is to exploit the spatially varying spectral, polarimetric, and temporal aspects of images. The insatiable demand for wide area coverage and increased resolution, as well as wide spectral coverage, is forcing the development of numerous sensor elements and systems across much of the electromagnetic spectrum. Both MEMS and nanotechnology will play a significant role in the overall systems that answer these needs.

Ultraviolet Light Sensors

Imaging for the ultraviolet portion of the spectrum is not as well developed as imaging in the visible and infrared regions. More efficient UV laser sources[96] and detectors that are solar blind are being developed for biodetection, missile threat warning, and UV communication.[97] These sensors are not covered in depth, as the findings and recommendations are not significantly different from those for the infrared, discussed below.

Visible Light Sensors

Most of the current sensing in the visible spectrum is performed using silicon semiconductor devices, either singly, in linear arrays for spectrometers, or as two-dimensional arrays in camera systems. Beyond the improvements in electronics discussed elsewhere, micro- and nanotechnologies are impacting visible spectrum sensing dramatically and will continue to do so over the next couple of decades. Recent advances in visible spectrum sensing have been primarily cost improvements in focal plane arrays. Current state-of-the-art uses for microtechnology include micro- and nanoscale particles in suspension for rapid polishing of lenses. Medium-term advances include mostly ancillary system improvements in antennas and electronics, especially jam resistance. The greatest improvements in visible sensors will come in the form of smart, multispectral sensor systems capable of robust target identification and able to withstand laser blinding and other countermeasures.[98]

Another area that may well provide value for the U.S. Air Force involves enhancing the visible light sensor system that comes with the human body, the eye. The visual function of eyes of pilots or other decision makers will eventually be modified for special purposes. Advances in vision correction[99] are leading to the ability to surgically alter a person's vision to suit specific needs, perhaps improving contrast sensitivity while trading depth perception for monochromatic vision,[100] as needed in IR viewing. It is possible to give the sniper or even a general crewman 20:10 vision. Successful long-term modification of the cornea of the human eye will require a better understanding of the long-term consequences of damage we cause to the functioning and physiology of the human eye as we perform invasive surgery like LASIK. Much of the physiology of the cornea is rooted in the ordered distribution of its 65-nm-diameter fibrils and fluid flow.

Infrared Sensors

Infrared sensors are used for night vision, for reconnaissance and surveillance from aircraft and space, and for homing of optically guided missiles. Current efforts aim to reduce the need for active cooling and to increase the spectral coverage. Space-based surveillance specifically requires improvements in power consumption, cost, and size in far-infrared sensors.[101] The ongoing revolution in nanomaterials may also provide better infrared-transmitting materials for windows and seeker domes.

Reduced cooling needs for IR detector arrays are being explored in several ways, including biomimetic sensors,[102] quantum dots and carbon nanotubes,[103] and micromachined bolometers[104] and bolometer arrays. Uncooled micromachined bolometer arrays now cost less than 10 percent of cooled military infrared images. Their size will increase and cost will plummet further as they are adopted for use in consumer electronics.

Advances in nanotechnology for monolayer-level control of epitaxial semiconductor layer growth are having a significant impact on cooled IR sensor systems. Improvements are being made to molecular beam epitaxial (MBE) precision growth of complex, multilayer IR detector arrays based on HgCdTe and related compounds in the difficult II-VI semiconductor family. Extraordinary advances in the capabilities of IR imaging systems, missile seekers, and space surveillance IR systems have been achieved and are continuing. Especially significant is the realization of multispectral IR sensors (requiring three or more photovoltaic junctions). Larger format, higher resolution IR imaging arrays based on the use of silicon substrates with II-VI IR sensor arrays, including the direct growth of high-quality HgCdTe on silicon, as well as HgCdTe on CdZnTe, are under development.[105] Integration with silicon technology also provides the potential for lower costs at higher volumes by adopting some of the silicon microelectronics industry's fabrication tools and approaches. However, this will remain an elusive goal until volumes of any one design are sufficient to warrant a concerted process development effort.

The development of novel avalanche photodetectors (APDs) and arrays of detectors is key to active Ladar (IR) sensor systems. These new devices make use of complex multilayer structures grown by MBE from II-VI or III-V materials. The II-VI material is usually HgCdTe or a related alloy. Similar novel devices using the InAs/GaSb-related family of III-Vs have also been reported.

The design and realization of entirely new types of IR sensing devices, such as the InAs/GaInSb "Type II" superlattice IR detector,[106,107,108] are clearly of interest to the Air Force. Proper operation of this particular device depends on the precise control and periodic replication of very thin (a few to tens of monolayers) alternating layers in a superlattice structure. Theoretical models for this type of detector indicate high IR sensitivity at substantially higher operating temperatures than in conventional photovoltaic devices. It is hoped that this will improve upon current detectors based on extrinsic silicon that must operate at below 20 K to reduce dark current.

RF Sensors

Nanotechnology has the potential to significantly improve the performance of RF sensors. These devices serve as the "eyes" of electronic systems that operate in the RF to millimeter-wave regions. Generally, these devices are nonlinear elements, such as diodes, that receive RF energy and convert it to electrical signals, which are then detected and further processed. Either direct detection or heterodyne detection techniques can be employed. Direct detection systems are simpler and easy to implement but have limited sensitivity. Heterodyne systems are more complex, requiring local oscillators, but have inherently greater sensitivity. Both systems require a nonlinear element in the front end as the interface between the RF and electronic environments. Although diodes are generally used, it is also possible to make use of active devices such as field-effect transistors, or high electron mobility transistors (HEMTs).

The fundamental limitation to the frequency performance of a rectifying front-end element can be described by the cutoff frequency, which can be expressed as

$$f_c = \frac{1}{2 p R_j C_j}$$

where R_j is the resistance of the diode junction and C_j is the diode capacitance. For optimum performance these devices are designed so that the maximum amount of RF energy is coupled into the diode junction area and directed across R_j. The diode capacitance is a parasitic parameter that acts to shunt RF energy around the diode junction, thereby degrading conversion efficiency. For this reason, it is desirable to minimize the device capacitance. This can be accomplished by scaling the diode area to very small dimensions. Diodes with cutoff frequencies in the terahertz regime can be produced. For example, GaAs Schottky barrier diodes with diameters of 0.15 micrometer have a cutoff frequency of 3.4 terahertz.[109] Nanotechnology scaling can push these cutoff frequencies to over 10 terahertz.

Nanotechnology offers the potential for significant improvements in RF sensor performance. Improvements will come from two directions. First, nanotechnology advances will result in improved technology for material growth that will permit ultrasmall devices to be produced with atomic-level control of the semiconductor layer thickness. For example, Schulman et al.[110] have demonstrated a Sb-heterostructure diode that can provide temperature-insensitive performance at frequencies exceeding W-band. These diodes were fabricated using an InAs/AlSb/GaAlSb heterostructure in a lattice matched configuration grown by MBE and had an area of 4 square micrometers. These devices can be used as backward and zero-bias diodes with both high sensitivity and low direct current power requirements.[111] RTD designs can push frequency performance well into the terahertz region and still maintain good detection sensitivity. Second, nanotechnology offers the potential to integrate intelligence into structures at the system front end. That is, a certain amount of processing can be integrated into the nonlinear element right at the point of RF to electronic conversion. At this location, the process of extracting useful information from the RF signal can be performed with high efficiency, thereby avoiding losses associated with transferring the signal further into the system. This paradigm would also permit simplification of the processing circuitry. Nanotechnology offers the potential to build intelligent processes into the semiconductor devices structure. These approaches are in their infancy but are likely to provide advances for numerous applications of interest to the Air Force.

Multispectral Sensing

Combining two wavelength regimes into one sensor housing conserves space and weight. Early attempts at this included simply applying a single InSb detector element on top of a single HgCdTe detector element. Beyond about 5.5 micrometers, the InSb element became transparent, allowing longer infrared wavelength to pass to the HgCdTe device. The advantage of this was an improved signal-to-noise ratio over the range of the InSb device (1-5.5 microns) and optimized HgCdTe detector element performance tuned to the longer wavelengths. The result was very good performance for 1-25 micron detection. A developing example of multispectral sensing is the combining of mid-wave IR (3-5 microns) detectors based on HgCdTe with wide bandgap III-V-based UV detectors along with silicon-based visible and near-IR detectors.

Sensors for Ultrasonics

Ultrasonic sensing uses MEMS devices for three-dimensional detection and tracking of sound. This work is aimed at detection and classification of underground facilities, ground vehicles, and missile launchers.[112]

Sensors for Physical Properties

Navigation Sensors. Spacecraft rely on sensors to determine their orientation with respect to the Sun (for power), Earth (for sensing and communication), and the stars. Many sensor types are used to determine this orientation, including optical sensors for the Sun and stars and magnetic field sensors for determining their orientation with respect to Earth. More recently, monitoring the phase shift in the signal from different GPS satellites has become useful.

Microoptoelectromechanical systems (MOEMS) can significantly decrease the mass, volume, and power requirements of optical navigation sensors, while MEMS could have a similar effect on inertial navigation sensors. A conceptual design for a small, lightweight, low-power, single-chip, micromachined, single-axis Sun sensor with 64 small interdigital detectors, is given in Figure 3-10. Coarse position is determined in digital mode by locating which element has the highest output; fine position is determined by ratioing the output powers from neighboring detectors.

Recent emphasis in the commercial market on E911, a concept consisting of various methods for providing geolocation services for stranded cell phone users, may lead to improved miniature GPS systems. As GPS systems shrink to fit into cell phones, the military counterparts can also be expected to shrink. The impact of tiny GPS systems on inertial measurement units (IMUs) and guidance systems in general will be tremendous.

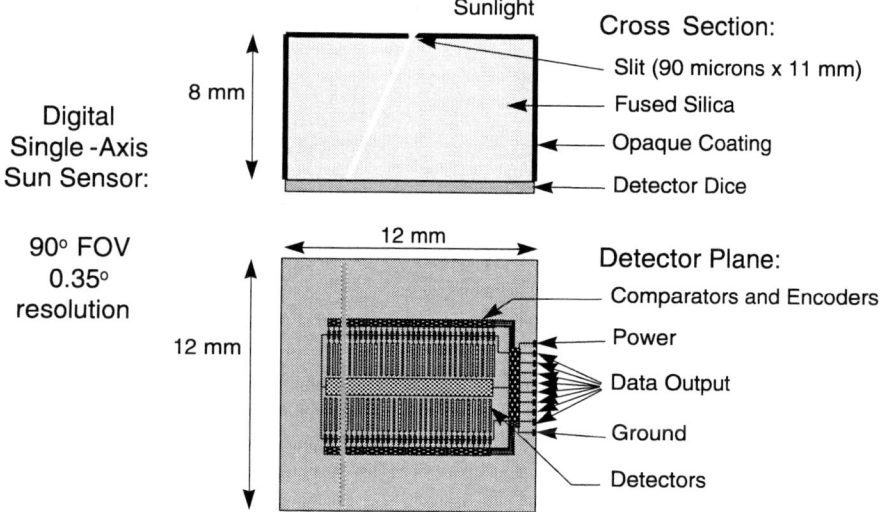

FIGURE 3-10 Micromachined Sun sensor. SOURCE: Helvajian, H., and S.W. Janson. 1999. Microengineering space systems. Pp. 29-72 in Microengineering Space Systems. H. Helvajian, ed. El Segundo, Calif.: The Aerospace Press. © The Aerospace Corporation, used by permission.

Inertial Sensors. Spacecraft IMUs use micromachined accelerometers,[113,114] gyroscopes, and other angular sensors to monitor their own forces and track their position and orientation. Gyroscopes are used to monitor spacecraft attitude, with optical sensors for absolute calibration. These gyroscopes and angular rate sensors are typically based on a rotating mass, a vibrating fork, or the circulation of light around a closed loop. Micromachined angle rate sensors monitor Coriolis forces, which are proportional to the angular rotation rate.

Inertial sensors consist primarily of accelerometers and gyroscopes. The current trend is to make these in silicon using MEMS, which allows dramatic reductions in weight, size, and power while also improving robustness. MEMS accelerometers are commonly used as impact sensors in automotive airbag systems.[115] New silicon micromachined microgyroscopes will soon be available. In addition, an even newer angle rate sensor design that uses quantum tunneling for the transduction mechanism is being tested.[116] This design may allow yet smaller (hence cheaper and lower power) angle rate sensors to be produced. Smaller angle rate sensors with accuracies similar to fiber-optic gyroscopes would allow more precise control of smaller vehicles, including arrays of nanosatellites, missiles, and UAVs.

When the IMU loses track of its star or the Sun, it must search rather broadly to regain lock. The addition of a very fast GPS could reduce the search area dramatically, providing a much faster lock. However, for reasons of weight and package, most IMUs do not have GPS capability. Micro GPS units, possibly drawing upon e911 cell phone technologies and MEMS, might play a part in bringing this capability into IMUs.

Magnetic Field Sensors. Low earth orbit (LEO) spacecraft typically use flux-gate magnetometers to measure local magnetic field strength and direction for control of orientation. Flux-gate, magnetoresistive, and Hall-effect sensors are all suitable for developing microengineered magnetometers. Honeywell has a three-axis magnetic sensor hybrid based on magnetoresistive transducers,[117] Nonvolatile Electronics, Inc., manufactures application-specific magnetic sensors based on the giant magnetoresistive (GMR) effect,[118] and another magnetometer concept is being developed at Johns Hopkins University[119] using the Lorentz force to measure vector magnetic fields.

Sensors for Chemical and Biological Agents

Chemical and biological sensors are of great importance, particularly in light of recent events. The Army, with much greater concentrations of troops than the Air Force, has taken a leading role in developing sensors for chemical and biological detection. Only a brief overview is presented here. A recent NRC report covered this topic in much more detail.[120]

Numerous approaches are being developed to detect chemical and biological attack on personnel, with nanoscience rapidly being recognized as a major contributor. Miniaturization of electronics and the concomitant ability to measure the behavior and properties of nanometer-size bits of matter introduce remarkable advances in sensing and detection. Microfabricated sensors provide a sensitive measure of nanometer (e.g., molecular) interactions with a high degree of specificity. These sensors are able to detect minute traces of chemical and biological agents as well as explosive vapors. Fabrication of sensor arrays and devices by the usual parallel processing methods, similar to chip fabrication, will mean inexpensive sensors for a wide variety of substances. MEMS production methods further expand the horizon of new devices used for these purposes. Such point sensors will be effective for environmental monitoring, security (buildings, transportation platforms, and points of embarkation), and battlefield protection. The impact of research in this area on mine detection could also be significant. This aspect of nanotechnology will probably lead to marketable products in the near future.[121]

The sensitivity of these sensors is due to the nature of the instrumentation developed to observe small changes at nanometer dimensions. Tools such as scanning local probes (tunneling microscopy and many related techniques) are just one of the many tools that may be used for sensing the presence of a specific molecular species. The fabrication of tools able to measure minuscule changes in mass, along with changes in optical, electric, and magnetic properties at nanometer dimensions, enables detecting the presence of submicron bits of matter held in place with the forces between single (or a few) molecules. With appropriately designed transduction, a single molecular unit (or nanoparticle) may be detected. A variety of methods for such measurements is becoming available. The issue then becomes how to bring the occasional nanoparticle into contact with the sensing unit. Such sensitivity is attracting great interest in light of the need to rapidly detect trace amounts of materials used not only on the battlefield but also by terrorists, such as anthrax, botulism, or ricin-toxins that are fatal in microgram doses.

Selectivity of these sensors is obtained through the forces that depend on the nature of molecular interactions and the method used for detection. By using biological specificity with polypeptide or DNA-type interactions, highly specific identification is possible. Antibody-antigen interactions result in highly selective binding at the nanometer scale. Attractive forces between molecular species vary, as evidenced through solubility or adsorption characteristics, and these forces may be quite specific for properly chosen molecular structures. By comparing the changes in response of an array of different materials exposed to an agent, a characteristic pattern can reveal the agent's identity.

Examples of the phenomena currently being considered for sensors illustrate a few of the approaches being considered. The mass of a nanostructure changes with events that bind additional nanostructures to it. Miniature oscillators, such as a quartz crystal microbalance, may be used to measure a resonant vibration frequency for a given structure. A vibrating cantilever is reported to be sensitive enough to detect the presence of a single *E. coli* bacterium in air with a mass of less than 1 nanogram.[122] The selectivity depends on the complementary nature of the two species interacting. It is estimated that the sensitivity can be improved by as much as four orders of magnitude. Another novel approach measures the amplitude of a transverse vibration as a function of increasing amplitude. When bonds between an adsorbed species and a surface rupture, the amplitude changes, enabling sensitive detection of selective components adhering to a surface.[123]

Changes in electrical conduction represent another approach to nanosensors. One such recent effort reveals changes in the conductivity of carbon nanotubes (CNTs) on exposure to various gases.[124] The conductivity of CNTs changes by three orders of magnitude in 2 to 10 seconds when 200 ppm of a gas such as NO_2 is introduced. The rapid response time is a direct result of the small dimensions of the CNTs; semiconductors used for this purpose typically take minutes to respond. The change in conductivity is probably due to a hole doping mechanism in the CNT induced by the presence of the NO_2 gas. An alternative approach that operates in solution involves nanopores in a membrane designed to bind complementary molecular units. Electrical conductivity through a nanopore continues until a complementary unit is attracted to the site and binds, blocking the passage of ions through the nanopore. The change in conductivity registers a selective binding event and the presence of a nanoparticle or large molecule.[125] An alternative approach uses small pores between electrodes coated with materials complementary to a desired molecular unit. An analyte that contains gold particles and a suspected DNA sequence will attach the gold between the electrodes, registering a significant change in the conductivity between the electrodes after suitable treatment. This scheme has been reported to give sub-picomolar sensitivities.[126]

Magnetic fields from submicron beads are used to detect forces between nanoparticles with the Bead Array Counter (BARC).[127] In this method, carefully designed target DNA materials are first patterned on top of a chip containing an array of micron-scale GMR strips. These strips are configured to electronically sense very small magnetic fields. When a solution containing DNA from unidentified pathogens flows over the corresponding area containing complementary target DNA on the chip, this pathogen DNA is captured on the surface. Micron-size magnetic beads, especially designed to bind only to the pathogen DNA molecules, are then injected into the solution and are bound to the selected site. The GMR sensors detect these beads, and the intensity and location of the signals indicate the concentration and identity of any pathogens present. The current BARC chip contains a 64-element sensor array; however, with recent advances in magnetoresistive technology developed for computer memory, chips with millions of GMR sensors will soon be commercially available. This advance will speed the development of a chip capable of screening for thousands of analytes simultaneously. Because each GMR sensor is capable of detecting a single magnetic bead, in theory, the BARC biosensor should be able to detect the presence of a single pathogen DNA molecule.

Optical indicators are successful sensors using well-known coalescence phenomena. Nanoparticle-based colorimetric detection is an effective sensing technology already in the marketplace for pregnancy detection. The principle involves antibodies attached to gold nanoparticles. The presence of an antigen or other complementary interacting agent binds the gold nanoparticles into clusters having modified optical properties (due to the larger cluster dimension), causing a change in color.[128] Surface plasmon detection, sensitive to local refractive index changes, provides an important technique for monitoring analyte adsorption.

Advances in chemical, biological, and explosives sensors are expected to be rapid, paralleling the increasing activity in nanotechnology in general. Bio-nanotechnology is even being studied as a means of destroying chemical and biological agents.

Self-Sensing

To ensure that the mission is completed, every part of the sensing system can and sometimes should be monitored for performance. In the case of manned aircraft, this includes the pilot, the crew, and their environment (e.g., CO_2 levels in the pilot's G-suit and cockpit). In the case of satellites, the environment of the payload, the structure, the launch site (hydrazine on the tarmac) and the payload itself are candidates for monitoring.

Load Monitoring

Load monitoring has historically been achieved primarily by individual strain gages (a typical F-18 has seven). However, owing to system weight penalties and the difficulties of installation and repair, they have been used infrequently in aircraft and space applications. Recent developments in optical fiber sensors[129] and wireless transponders[130,131] have helped solve some of these difficulties. However, there has not been much successful commercialization of these technologies, primarily because of limitations on the availability of power and very small antenna elements, both of which may have micro- and nanotechnology solutions, and overall lack of maturity of the embedment process and its associated egress and embedding issues.

Air Pressure Monitoring

MEMS-based flow monitoring has been pursued for underwater vehicles, but the most notable effort in monitoring the local environment of an aerostructure was developed by the Boeing Company and Endevco. Their pressure belt[132] (Figure 3-11) contains silicon-based multichip modules (MCMs), each of which contains a MEMS pressure sensor, a temperature sensor, and data acquisition and

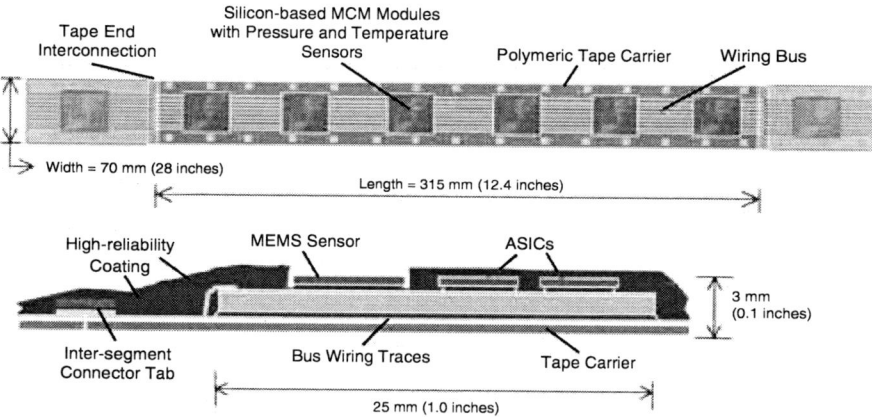

FIGURE 3-11 Boeing/Endevco pressure belt. SOURCE: North Atlantic Treaty Organization Advisory Group for Aerospace Research & Development (NATO/AGARD). 1996. Smart Structures and Materials: Implications for Military Aircraft of New Generation, AGARD-LS-205. Ottawa, Canada: Canadian Aeronautics and Space Institute.

MAJOR AREAS OF OPPORTUNITY 93

signal-conditioning circuitry on a flexible belt. Installed on an aircraft's wing, the pressure belt provides data during the aerostructure's design stage.

While the MEMS pressure belt described above is intended for measuring the pressure distribution on a wing in flight, nanoscience has provided an alternative solution for wind tunnel measurements. When fluorescent dye molecules are applied as paint they act as tiny oxygen probes, reacting by altering the fluorescent intensity in response to the local oxygen concentration. This pressure-sensitive paint (PSP) is illuminated with light to excite the fluorescence. As the pressure of the airflow across the surface changes, the intensity of the light generated is monitored, as shown in Figure 3-12, clearly indicating pressure gradients. The temperature sensitivity of the thin-film coating is corrected using infrared measurement.

Condition-Based and Prognostic Health Monitoring

Status sensing is done for both routine operation of systems—for example, as part of control loops—and for determining when routine or critical mainte-

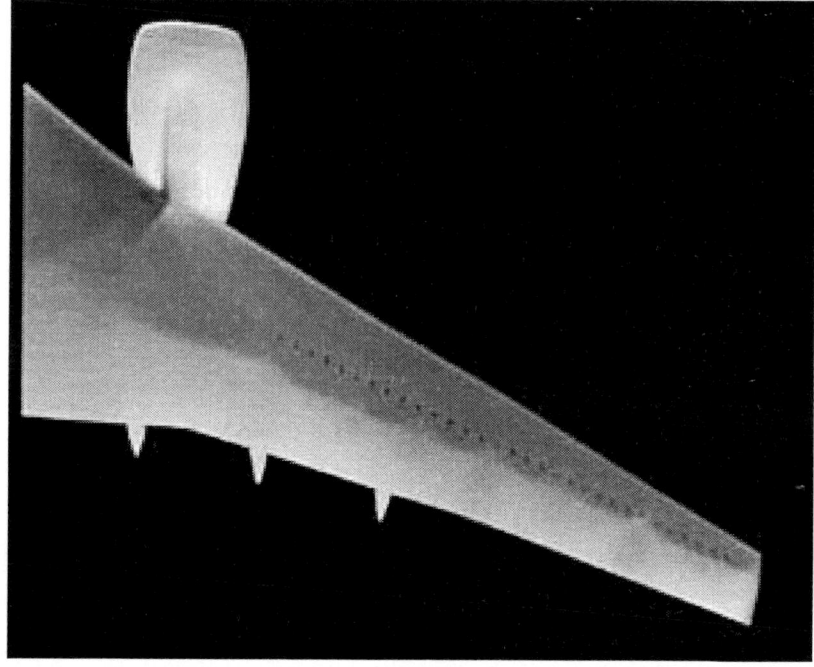

FIGURE 3-12 A typical pressure-sensitive paint result for a wind tunnel model of a transonic transport airplane. Courtesy of Boeing.

nance is needed. The components here are (1) the system being monitored and (2) the sensor system that is providing data on the status of the system of interest.

The list of functional systems of interest to the military, and to many industries, is long. Power generation, distribution, and use systems are many and varied. The same can be said of communications systems. Weapon systems are also remarkably diverse. The numerous installations and mobile platforms in and on which various systems are used further complicate monitoring possibilities. It would be necessary to make a list of systems of interest for monitoring in order to define a sensor architecture.

Sensors require a nontrivial set of related hardware and software. Smart sensors capable of turning data into information at the system, logging and analyzing it, and forwarding routine and emergency information require micro-controllers as well as associated electronics such as amplifiers. The entire sensor system needs power, the provision of which is usually the limiting factor in the lifetime and performance of a sensor system. Wired or wireless communications are needed. Finally, the entire system must be packaged in a housing.

The hardware items will have different susceptibilities to the employment of micro- and nanotechnologies. Sensors are first-rate candidates for exploitation of new technologies. Advances in microcontrollers and other ICs are driven by market forces in the semiconductor industry. Nanomaterials should have an impact on batteries and, possibly, solar cells.

Prognostic Health Monitoring from Satellites and On-Orbit Manned Vehicles

Many space-system operations occur on the ground, and micro- and nanotechnologies will probably be inserted into these operations before they are inserted into space operations as a result of the more benign environmental and reliability constraints. Miniaturized, multiparameter, MEMS-based sensors with integrated data loggers and wireless (optical or RF) communications will become important in production and ground operation. Relatively low bandwidth devices on the market that have peak-sensing capabilities (a.k.a., telltales) to sense transportation or to handle stress variables can ensure that maximum limits have not been exceeded during production, transportation, and storage operations. MEMS sensors for these parameters already exist and can be mass-produced for inexpensive environmental monitoring packs. Knowing what, when, where, and how a limit was exceeded is critical during the spacecraft flight readiness review.

Micro- and nanotechnologies can also be used to instrument launch vehicles. Current launch vehicles like the Titan IV are often instrumented to measure the liftoff and ascent flight environments. However, these vehicles often have a limited number of channels to characterize the dynamic acoustic and vibration environments. By proliferating sensor units on the launch vehicle, a better characterization of the environment is possible. Similarly, there is a need to instrument the launch site. Ground-based measurement of rocket ignition overpressure and toxic chemical release (e.g., HCl from a solid booster) are needed in conjunction with the launch vehicle monitoring system to dramatically increase the awareness of vehicle status and the launch site environment. MEMS sensors coupled to data transceivers can be used in a wireless network system onboard the vehicle and on the launch site. The telemetry data can channel real-time or near-real-time information to a ground-based data storage system for postlaunch review.

Another important role for micro- and nanotechnology in both satellite and on-orbit manned vehicle operations is in condition-based maintenance (CBM) status and health monitoring systems. These systems could save future costs by fault detection and isolation and by enabling automated self-test and repair actions to take place. The CBM protocol also enables safer operations as well as increased system availability in contrast to a failure-based maintenance protocol scheme. Reuseable launch vehicles are prime candidates for CBM. One relatively common malfunction in spacecraft is the failure of a high-speed bearing, reaction wheel (i.e., momentum wheel), or gyro bearing. Bearing degradation can often be anticipated by monitoring vibration signatures; this would be an excellent application for a micromachined accelerometer coupled to a digital signal processor or microprocessor in an application-specific integrated microinstrument (ASIM). Corrective action could consist of the metered release of lubricant via a fluidic ASIM. Smart bearings, smart structures, and multifunction structures are already being considered by space engineers. These ideas have also been considered in the aerospace community for developing adaptive structures that have embedded sensors, actuators, controllers, and processors.

Distributed Sensor Systems

Swarms of sensors can provide additional value by simply providing sufficient coverage. One obvious example is dense sensor arrays for space weather forecasting.[133] However, swarms of sensor suites can also provide a more complete information space for other purposes, including building a spaceborne equivalent of the Very Large Array (VLA) (Figure 3-13). The VLA consists of 27 radio antennas in a Y-shaped pattern just west of Socorro, New Mexico. A spaceborne VLA could be sized to aim not out into space but at Earth's surface.

FIGURE 3-13 The Very Large Array. SOURCE: Photo by Dave Finley, NRAO/AUI.

An array of coherent, precisely controlled cameras could dramatically increase resolution. The data from VLA are combined to give the resolution of an equivalent antenna 22 miles across, but the array has the sensitivity of a single dish 422 feet in diameter.

How does one achieve this coherent, ordered array of cameras? Orienting and positioning perhaps hundreds of cameras, each perhaps miles apart, would require tremendously accurate relative positioning, which, in turn, requires very fine timing units. While timing comes under the purely electronic aspect of sensors, which is generally not covered in this chapter, a short mention of work in this area might be of benefit to the reader owing to the scarcity of information in the literature. Recent work on terahertz timing has been aimed at what is now called a chip-scale atomic clock.[134] An alternative approach for achieving this timing is being explored wherein a 10-megahertz spacing frequency comb extending across more than an octave in the visible is achieved by spectral broadening of a femtosecond laser pulse train in a photonic crystal fiber.[135] The upper limit of the uncertainty of this approach has been measured to be less than 5.1×10^{-16}. For use with VLA nanosatellites, this laser clock would have to be miniaturized, the province of MOEMS. One revolutionary implication of having large arrays of spaceborne sensors is an operational Discovery II vehicle, a JSTARS in space. Twenty-four satellites could cover instantaneously one-eighth of the planet's surface and track every vehicle.

Sensors composed of smart dust would also qualify as distributed sensors. These devices are distributed to the wind in an area of interest and perform their task without actively controlling their arrangement. Smart dust networks[136,137] are massively distributed sensor networks, consisting of hundreds to thousands of autonomous sensor system nodes and interrogators to query the network. The enabling technologies for these devices come predominantly from MEMS.

How does the Air Force assimilate all of the data provided by myriad arrays of sensors from all parts of the spectrum, especially when it is combined with data from weather reports, troop movements, telltale sensors, news reports, and other sources of intelligence, and then display it in a usable manner? Data fusion is the field that addresses appropriately combining massive quantities of data. Micro- and nanotechnologies will play a role in data fusion by providing faster and smaller high-performance computing and information processing.

An important concept arising from systems designs for large numbers of dispersed sensors acting in coordination is that of emergent behavior. This refers to new capabilities and characteristics of the group that emerge from the cooperation between elements and that are not present for single or small numbers of sensors. This concept has numerous biological analogs (see Box 3-2).

Swarms of sensors or platforms are enabled by embedded intelligence and communications channels. Sensor networks can be relatively simple if they respond only to preset stimuli, do not move, and do not accept commands once activated. Reprogrammable devices such as Flash RAM and field-programmable

BOX 3-2
Emergent Behavior of Swarms of Microplatforms

A swarm is a collection of a large number of relatively simple components. Ants in a colony, termites in a mound, and bees in a hive are examples provided by nature. Each individual has limited capabilities, but the group can perform large-scale feats that ensure its survival. The swarm, not the individual ant, termite, or bee, is the important entity. The swarm can do things that go far beyond the capabilities of the individuals. Foremost, the swarm exhibits behaviors and performance that "emerge" from the individual activities but are vastly more complex. Also, the group can function for many times the lifetime of the individual.

Swarm intelligence is an active area in science that has yet to become technologically useful. In fields ranging from etymology to computer science, the study of actual and simulated swarms is of great interest. The growing capabilities of microelectronics, -mechanics, and -optics may enable the practical use of the collective emergent behavior of many interacting miniature systems by the military. Numerous small but cheap and sufficiently capable terrestrial robots could be of use to the Army and the Marine Corps, especially in urban warfare. All Services might benefit from the availability and behavior of large numbers of small air vehicles that are enabled by microtechnologies. Large arrays of picosatellites also offer the possibility of overall performance that is substantially greater than the product of their individual abilities and their number (Figure 3-2-1). In all cases, the tight integration of sensing, signal-processing, computation, and communication functions that is possible because of parallel mass fabrication of microsystems increases performance and keeps down unit costs. Future sensor systems-on-a-chip will be required for the cost-effective production of swarms of useful microplatforms.

FIGURE 3-2-1 Swarm of nanosatellites. SOURCE: Lewis, D., S.W. Janson, R.B. Cohen, and E.K. Antonsson. 1999. Digital micropropulsion. Pp. 517–522 in Proceedings of the 12th IEEE International Conference on Micro Electro Mechanical Systems (MEMS '99). Piscataway, N.J.: Institute of Electrical and Electronics Engineers, Inc. © 1999 IEEE.

gate arrays enable real-time reprogramming of sensor networks to respond to different stimuli, to self-configure communications channels or sensor gains for the current environment, and to adaptively coordinate data transfer to a given user as the RF and power environment change. This programmable flexibility should allow mass production of standardized, highly integrated systems that can be used in a wide variety of environments for different applications.

Initially, these systems will be user-preprogrammed before deployment. As the numbers of individual units grow into the thousands for a single application, custom programming of each unit will become excessively tedious. Some degree of intelligent self-configuration and autonomous system optimization will be required. More autonomy, such as adaptive coordination, can be included in new systems as the theoretical modeling of system behavior and operational experience improves. An enhanced understanding of distributed adaptive systems will be required as these systems become smarter and self-reprogrammability improves. A challenge to the future warfighter will be to maintain enough control over distributed self-adaptive systems to ensure that they perform the desired function.

Finally, consider the additional complexity of having intelligent distributed platforms that move on the ground, in the air, or in space. Geometric configuration can change on demand to send groups of units to individual targets, to avoid mechanical or electromagnetic attack, to provide variable effective aperture for radio frequency reception or transmission, or to maintain system functionality as individual units fail or are eliminated. Techniques for distributed real-time optimization for specific functions with changing system geometry and network connectivity will have to be developed.

Findings and Recommendations

Finding 3-3. *In sensors, especially for remote sensing applications of importance to the Air Force, there are a large number of high-performance requirements and military-specific functions that are clearly beyond the scope of commercial interests.*

Recommendation 3-3. *The Air Force should support in a sustained manner the research, development, and manufacturing infrastructure for military sensor systems.*

Finding 3-4. *The resurgence of the commercial MEMS sensor industry represents an opportunity for the U.S. Air Force to harvest improvements in military-specific MEMS devices.* The maturing of the overall MEMS industry and increased MEMS activity in the telecommunications market have dramatically increased the performance of design tools and the availability of complex space-qualified MEMS devices.

Recommendation 3-4. *The Air Force should actively pursue improvements and/or adaptations of commercial efforts for military needs and be prepared for a larger investment at the 6.2 through 6.4 levels.* This recommendation is not meant to imply total reliance on commercial sources. Commercial interests may drive the market in a different direction than is needed by the military, and commercial interests may interfere with military interests in the MEMS-based community.

Finding T4. *Large, distributed fixed arrays and moving swarms of multispectral, multifunctional sensors will be made possible by emerging micro- and nanotechnology, and these will lead to significant fundamental changes in sensing architectures.* Concepts such as smart dust and distributed communication networks actively exploit the technological capabilities of emerging micro- and nanotechnologies. The fusion of data from large numbers of sensors as well as large numbers of sensor types will drive research in new networking concepts.

Recommendation T4. *The Air Force should develop balanced research strategies for not only the hardware but also the requisite software and software architectures for fixed arrays and moving swarms of multispectral, multifunctional sensors.*

BIOLOGICALLY INSPIRED MATERIALS AND SYSTEMS

Most natural materials exhibit a combination of desirable properties—for example, strength, flexibility, and light weight—that are not usually found in synthetic systems. Examples include spider silk and a bird's bone structure. The unique properties and performance of these natural substances arise from a precise hierarchical organization at the micro- and nanoscales, coupled with nature's use of composite materials incorporating inorganic components alongside biological materials, e.g., bone or tooth enamel. Therefore, it is important to investigate, understand, and ultimately reproduce nature's ability to precisely integrate organization at the atomic and molecular levels. Using the molecular forces of nature to create materials templates and/or to solve nanoscale assembly problems, e.g., self-assembled monolayers,[138,139] nanotubes, molecular motors,[140] biotin-avidin binding,[141] and hydrophobic/hydrophilic surfaces,[142,143,144] has already been demonstrated. Bioorganisms may one day be employed to perform directed nanomanufacturing of complex structures. Already, genetically engineered bacteria and plants are in use to produce drugs and other chemical compounds. There are biological organisms, e.g., radiolarians, that manufacture silica exoskeletons with highly complex geometric shapes. Given the rapid emergence of genetic engineering, it may soon be feasible to bioengineer radiolarians that can manufacture specific shapes on the surface of silicon chips to create the two-dimensional templates for integrated circuit fabrication, a task now accomplished with photolithography. Similarly, such organisms could be used to manufacture three-dimensional micromechanical scaffolding for MEMS and NEMS (nano-electromechanical systems). Alternatively, it may be possible to synthetically re-create their handiwork. Mesoporous silica has already been implemented in forming self-assembled, three-dimensional, porous layers with porosity that is controlled by the wavelength of light to which it is exposed.[145,146]

The interconnection between specific types of circuit elements, e.g., transistor gates, may also be achieved by self-assembling biosynthesized materials that are attracted by specific mechanical or chemical signatures to specific sites. Biological systems utilize highly elaborate and often dynamic self-assembled interconnection. Neurons are known to emanate from one type of cell and to search out, find, and attach to another cell by some sort of (presumably) chemical sensing mechanism. This neural wiring occurs, for example, during all stages of development, and during recovery from neural injury. Guidance of cell growth by grooves etched in silicon surfaces[147] was a recent demonstration of directed self-assembly using microscopic topographic effects.

Biomimetics for Improved Sensing, Communications, and Signal Processing

New sensing, communications, and electronic signal processing ideas might be drawn from an understanding of the "intelligent" behavior of neurons and cells. Cells are known to sense and respond to thermal, optical, chemical, mechanical, and electrical stimuli, e.g., elongation due to shear stress,[148,149,150] and galvanotaxis,[151,152] but the underlying mechanisms are not well understood. For example, what is the electric field sensor in an epithelial cell or the shear sensor in an endothelial cell? Communication between adjacent cells is also evident but largely unexplained. In large neural networks, the mechanisms by which the frequency of oscillation is produced and controlled and by which the phase relationships among oscillatory neurons are maintained are generally not understood. Yet, oscillatory dynamics within neural networks, such as those associated with simple rhythmic behaviors, motor control, sensory perception, and sleep, are found at all levels of the nervous system in both invertebrate and vertebrates. For example, central pattern generators, which control rhythmic motor behaviors,[153] are networks of neurons which by themselves can generate a motor pattern associated with a particular behavior, in the absence of sensory feedback or descending control from higher centers. The study and understanding of these and similar biological networks[154] may produce new directions for communications theory and practice.

As our understanding of cell membranes expands, the idea of creating artificial cell membranes as sensing elements will become a reality. Bilipid membranes with selective transport elements, e.g., ion channels, that respond by opening or closing in response to a target biotoxin have already been demonstrated.[155,156] This device is a synthetic mimic of a nerve-cell membrane. The very large difference in open versus closed ion-channel conduction provides built-in amplification for this sensing mechanism, promising single-molecule detection sensitivity. Integration of this technology with integrated circuits is an obvious next step to provide the necessary high-density signal acquisition and processing.

Biocomputing is a new term that refers to the use of highly specific binding of molecules in sequence, e.g., the hybridization of DNA molecules, to perform mathematical or computing functions. Modest examples of biocomputing have already been demonstrated.[157] However, since biological systems are less than perfect, a synthetic system based on the same scheme is envisioned to achieve reliability. Nature relies on imperfections to evolve and on redundancy and adaptation to survive. Synthetic systems design could also benefit from these features, which could allow them to achieve autonomy, self-repair, and adaptation to new environments or situations.

Biological macromolecules, cells, and sensory organ systems utilize energy transduction mechanisms that often are not primarily electronic in nature. The primary response to a stimulus often begins with a chemical release, conformational change, or mechanical deformation, followed by secondary transduction events. MEMS and NEMS now enable the use of mechanical reaction to stimuli as a means of sensing. Examples include microbridges with chemabsorbing layers[158] that change mass, which in turn changes their resonant frequency, and microcantilevers that deflect when DNA hybridization takes place on their surface.[159,160] Some MEMS sensors have been demonstrated that mimic the mechanical transduction of specific biological sensory organs, including a directional acoustic resonator based on the fly's ear[161] and an accelerometer with fluid mass[162] based on the human inner ear.

Many animals possess exceptional sensing capabilities for which artificial replicas have yet to be demonstrated—for example, the heat-sensing capabilities of snakes, the nighttime visual acuity of owls, pheromone detection by moths, odor detection by kiwi birds and dogs, and vibration detection by spiders. New and improved methods of detection, location, tracking, and identification of human life signs and ground-based vehicles are needed for the Air Force as well as other branches of the military. Currently, much focus is placed on electromagnetic means, e.g., infrared detection. However, improved sensitivity to specific gases and molecules, e.g., CO_2, pheromones, and NO, would enhance the ability to detect the presence of enemy troops. Vibration detection and identification through frequency signatures could be used to detect and track ground vehicles. It has been observed that many animals can detect approaching or faint earthquake tremors before humans and seismic instruments. What is being sensed and how are yet to be discovered. Reconnaissance mission capability could be expanded by using drones to deliver gas- and vibration-sensing systems to ground locations, where they autonomously set up and begin intercommunication and then proceed to perform, for example, triangulating methods of detection and tracking and to transmit the information to airborne vehicles.

The actuation mechanisms and signal generation capabilities of biological systems also have relevance to the development of autonomous microsystems. Bioluminenscence of fireflies and many fish, electric-shock-generating organs of eels, water-powered actuators of jellyfish, and the jumping ability of fleas are all examples of highly developed biological actuators. Applications to micro- and nanosystems are many, including energy conversion for power generation, mechanical amplification, and optical readout.

Enhanced Human Performance—The Machine as Part of the Man

Micro- and nanotechnology offer new inspiration for an old concept, the bionic man (or cyborg). Implanted sensors, neural interfaces, and muscle-controlling devices have appeared in science fiction novels and movies for many decades. In recent years, biomedical engineering has made numerous, significant advances in this arena, including the artificial heart, artificial joints and limbs, and muscle-stimulating electrodes. Micro- and nanotechnologies make these devices less invasive by reducing size and increasing "smarts" through the integration of high-speed signal processing and control circuitry and through advances in biocompatible materials. These advances reduce the problems associated with implant surgery and rejection and improve the lifetime, capability, and performance of biointerfacing devices. For the Air Force, improved human performance could mean better vision, enhanced by IR detectors interfaced to the optical cortex, or monitoring of pilot brain function to intervene in the case of high-G blackout or falling asleep during any machine operation. Another application is in self-diagnosis, medication, and accelerated healing of personnel in the field. Intelligent bandages that use electrical stimulation to accelerate healing and detect and treat bacterial infections are currently in the research stages. Microneedles for drug delivery and for biofluid sampling are in product development today. It appears that the "hypospray" and "tricorder" of Star Trek fame are not that far away from reality. Other enhancements of human performance may come from a reduction in sleep and food requirements. Transdermal feeding and regeneration devices are quite possibly next-generation developments of microneedles and electronic bandages. Wearable computers and electronic clothing that monitor the wearer's vital signs are already in development. Enhanced capabilities for these "garments" are foreseen—for example, electronic camouflage, body temperature control, hydration control, and proximity and environmental monitoring are all feasible. The Air Force could apply self-reliance advances to enhance the survival of downed pilots and prolong the endurance of personnel in space. In all instances involving implanted devices, research will require FDA approval as well as the federally required oversight by the local institutional review board as a safeguard for any research involving human subjects.

Findings and Recommendations

Finding T3. *Biological science offers new opportunities in nanotechnology systems, especially for sensors, materials, communications, computing, intelligent systems, human performance, and self-reliance.* Millions of years of evolution have produced highly specialized sensing and communication capabilities in nature. Understanding of how these sensors work is growing but is still very limited. As the fundamental mechanisms are discovered and studied, applications rapidly follow. Advances in micro- and nanotechnology have enabled discovery in biological systems, which in turn has provided new means of sensing and communicating. Clearly, advances in technology and in the biological sciences go hand in hand in developing new capabilities.

Recommendation T3. *The Air Force should closely monitor the biological sciences for new discoveries and selectively invest in those that show a potential for making revolutionary advances or realizing new capabilities in Air Force-specific areas.*

STRUCTURAL MATERIALS

Introduction

The range of operational requirements for the most visible parts of military systems—the physical structures and platforms—is exceptionally broad and often extreme. Structures of special military importance include satellites and other spacecraft along with their specialized structural components, aircraft, land vehicles, water vehicles, missile systems and other weaponry, and warfighter support and protective equipment. In general these structures and platforms need to be lightweight; exceptionally strong, tough, and durable; tolerant of extreme temperatures (especially very high temperatures); and suitable for use in extraordinary environments such as high altitudes, space, saltwater, desert, arctic, and other extreme climates.

Future systems may be given enhanced capability by endowing structures with multifunctional attributes. In addition to housing and protecting its contents, a structure may be required to alter shape, actively provide cooling, reduce flammability, sense its environment, repair itself or change its dielectric or other properties.

A structural materials topic that is especially relevant to this study but generally overlooked when considering nanoscience's importance to macromilitary systems is structural materials for microelectromechanical devices. Since the miniaturization of electronic and sensor systems by way of MEMS technologies is essential for future defense systems, it is important to consider the technical issues and opportunities associated with the nanoscale materials needed to construct MEMS components.

Four areas projected to have major impacts on emerging nanoscale structural materials for Air Force systems and missions are discussed in this chapter. These areas are the following:

- *Lightweight materials.* Tough, strong, and durable lightweight structures and structural materials would enable lighter and faster aircraft and weaponry, with lower associated logistics costs. Such technology would also allow the launching of satellites and other spacecraft at less cost both because the structures on the satellites themselves are lighter and because subsystems used in the launch, like hydrogen fuel tanks, could be made more durable, more resistant to damage, and reusable.
- *Improved coatings.* Wear-resistant and corrosion-resistant surface coatings and tribological surface treatments would improve the durability and function of structure surfaces. Improvements in surface coatings and treatments for aircraft can reduce maintenance costs and extend the useful life of airplanes and other systems.
- *Multifunctional structures.* Multifunctional structures combine other functional features with purely structural uses such as thermal management functions, sensing, movement or shape change, energy storage, and self-inspection and self-repair. Such capabilities would improve thermal designs for supersonic aircraft and missiles and for spacecraft entering or leaving Earth's atmosphere, augment systems sensing capabilities, improve power and guidance systems, or simplify repair and maintenance processes.
- *Materials for MEMS.* Microstructural materials with reduced fatigue or creep and surfaces with suitable friction, stiction, and wear properties are essential to the successful transitioning of MEMS device technologies into their target applications. MEMS technologies will revolutionize Air Force system and subsystem designs by allowing the radical miniaturization of electronic, optical, mechanical, and sensing components.

Lightweight Materials

The quest to develop lighter structural materials with all the necessary mechanical properties normally found only in heavier materials (conventional metals, ceramics, etc.) is being pursued on several fronts. These include the strengthening of conventional materials, the development of lighter metal alloys, and the creation of novel composites or hybrids of a number of different material types combined to create a structural material by design.

Long before the term "nanotechnology" became commonplace, the benefits of manipulating the grain structure of materials on the nanoscale were known and exploited. Increased strength and reduced creep (or other irreversible, detrimental mechanical property changes over time) in finely grained, nanostructured materials is not well established in many polycrystalline materials systems. These nanoscale materials can provide for stronger, more durable, and more stable structures. The classic model for how the strength of metals increases as grain size decreases[163] describes the pileup of dislocations at grain boundaries, producing stress, which when added to applied stress results in slippage across the boundary. Smaller grains result in a smaller pileup of dislocations and less stress, with a larger external force needed to create slippage (and therefore a stronger material).

The goal of developing manufacturing methods and processes that produce materials with ever-smaller grain size has been pursued in metallurgy for decades. Nanoscale control of the grain structure of lightweight aluminum and aluminum alloys could be of value especially for aerospace structures. In addition, improvements in mechanical properties and the allowed operating temperature of titanium and titanium alloys are being pursued using nanoscience approaches. Large increases in the yield strength of metal alloys are well documented for nanoscale-grain materials relative to conventional microscale grain alloys. Methods of refining the grain structure of metal to the 50-nanometer (or so) scale have been reported.[164] In addition to greater yield strengths, combinations of desirable features such as improved ductility and strength can be designed in, as can hardness for metals and metal alloys composed of nanostructured materials. However, it can be difficult to avoid grain growth when metals with such fine grain sizes are subsequently heated. Researchers at AFRL and elsewhere are working on methods to increase the thermal stability of nanophase metals and alloys so that they retain their superior mechanical properties at elevated temperatures.

For ceramic materials, a goal has been to enable net-shape manufacturing of consolidated ceramic nanoparticles (such as titania or alumina) with a shape and dimensional precision beyond that possible with conventional processes. The role of nanoscience here is to aid in understanding the behavior of the nanosize grain boundaries, which can slide under stress without breaking bonds as a result diffusional healing, an atom transport mechanism that can take place over very short distances characteristic of nanoparticle-formed grains.[165] Ceramic structures can be made more easily and much less expensively if such net shape-forming processes are made more capable.

The subject of structural nanomaterials was given a comprehensive review at a recent Army-sponsored NRC Workshop on June 20-21, 2001.[166] The workshop dealt with the synthesis, assembly, processing, fabrication, and manufacturing of a variety of nanomaterials along with structure characterization and the modeling, simulation, and application of these materials. Methods for producing grain sizes of less than 50 nanometers for ceramics, metals, polymers, and composites were discussed. Most of the workshop participants, dealing with nanosize particles as basic building blocks, concluded that while progress is being made, challenges remain on virtually every front in developing processes to consolidate nanosize powders into useful and stable forms at useful manufacturing scales. Also discussed at the workshop were severe plastic deformation processes that "work" materials to produce ultrafine grain sizes and increased strength. The equal-channel angular pressing type of severe plastic deformation is reported to be especially promising for Air Force applications such as lightweight aerospace structures.[167]

The mechanical strength of nanosize, superstrong fibers such as carbon nanotubes and perhaps other materials such as boron nitride nanofibers, coupled with the lower density of these materials, offers the potential for much lighter composite structures that are stronger and tougher than conventional structural materials. This is a key advantage sought in the aerospace world for spacecraft, aircraft, and military systems. Theoretical studies suggest that the Young's modulus and the breaking strength of single-wall carbon nanotubes, for example, should be exceptionally high. Recent experimental measurement of breaking strength and Young's modulus tends to support the theory.[168] However, the mechanical properties of such nanowires, -tubes, or -ribbons must be not only probed and characterized but also manipulated and optimized at the scale of individual tubes and assemblies of tubes if these concepts are to be made practical.

The mechanical properties of assemblies of these wires or fibers must be understood as must be their properties in conjunction with the matrix material with which they are to be hybridized. Carbon fibers with a tensile strength up to 6 GPa are commercially available. Initial experimental measurements on 4-micron-long, single-wall carbon nanotube (SWNT) "ropes" consisting of tens to hundreds of individual SWNTs bound in van der Waals contact have yielded values up to 45 GPa.[169] The hope is that millimeter-long SWNTs can be formed into longer fibers or dispersed into a composite matrix while still maintaining a significant fraction of this observed order-of-magnitude improvement in strength over conventional carbon fibers.

Very large yield strengths (a measure of how much the material can stretch without breaking) of 5 to 10 percent have been observed for carbon nanotubes. This perhaps excessive ability to stretch may make this type of material impractical for aerodynamic structures, or, alternatively, it might be looked upon as an opportunity to incorporate a morphing material to produce something akin to flapping wings.

Methods for synthesizing nanotubes, controlling their orientation and bonding character, and producing them in macroscopic quantities are at an early stage of development. Also at an early stage are methods for fully characterizing the basic properties, especially the mechanical properties, of assemblies of structures,[170] and for using them in practical structures.

The superior aerodynamic performance enabled by use of lightweight structural materials of suitable yield strengths would be of particular value to the Air Force. Improvements in strength-to-weight and stiffness-to-weight ratios are sought. Subsonic aerodynamic performance at all altitudes improves with wing aspect ratio (span to chord ratio); this ratio is constrained in many applications by the structural weight penalties associated with very long, thin wings. Nanoscience innovations such as carbon nanotube composites offer the possibility of reducing the weight penalties sufficiently to allow aerodynamic performance to dominate the design.

Improved Coatings

Hard, durable surfaces with suitable aerodynamic features are essential to all aircraft, spacecraft, and other vehicles. The tribological properties of coatings are the key to wear resistance of radomes, windows, and many other components of military systems used in extreme conditions. It has been suggested by numerous studies that the design of super-hard, wear-resistant composite coatings is feasible using nanocrystalline constituents having the mechanical properties associated with the fine-grain structure such as resistance to classical dislocation formation and slip.[171] Coatings that can withstand extreme temperatures, abrasion, and wear are especially important for advanced aircraft and space vehicles.

For materials used as coatings or surface treatments, including composite materials, multilayer structures, and materials with unusual elastic properties, the relationship between grain size and orientation and material mechanical properties is complex. In general, the crucial mechanical properties of material structures are controlled by the size and nature of grain boundaries, and those properties may get better as the constituent particles comprising the structure get smaller. At sufficiently small grain size (smaller than several tens of nanometers), the classical relations may break down and more complex behavior take over, such as saturation of strength increase at levels well below the theoretical yield stress maximum derived from basic mechanical properties such as shear modulus. The real benefits that can be derived from implementation of nanostructured materials will therefore depend on a detailed understanding of controlling mechanisms (grain boundary formation; dislocation formation, annihilation, pinning, and movement) at the nanoscale regime.

A major practical limitation on the implementation of hard coatings, generally, is that there is inevitably residual stress in the films, which limits the thickness that can be applied and therefore the overall protective capacity of the coating. Some techniques are being explored to allow deposition of coatings that have grain sizes less than 100 nanometers. This not only can increase the hardness and toughness, as noted before, but also can dramatically reduce the residual stress state of the nanostructured coatings and increase by a factor of as much as four the coating thickness that can be applied.[172] Such thick, hard protective coatings would clearly be a substantial advantage for Air Force systems.

Nanotechnology is also employed in multiple dimensions in searching for ways to improve surface coatings. In addition to controlling the grain size of the basic materials in order to improve mechanical properties, novel schemes are being explored to use complex multilayers that are themselves constructed on the nanoscale at a few monolayers in thickness. In this regime, the mechanical properties of the composite films do not follow the same deformation mechanisms as micron-scale materials. For example, periodically structured multilayers of selected metal layers that are relatively soft on the macroscale can exhibit ultrahigh strengths in nanostructured layers.[173] Moving to the regime of interface- rather than bulk-dominated mechanical properties as we move to the nanoscale from the microscale is the key to constructing durable, hard coatings and thermal barriers on the surface of structures.

Multifunctional Structures

Incorporation of active material layers, fibers, or particles to produce movement, exert force, or permit alteration of mechanical shape under stimulus (e.g., for self-repair, or for deployment of mechanical structures) is an area just at the beginning of exploration. The development and application of active or smart materials (such as shape memory alloys and polymers, piezoelectrics, electroelastomers, and magnetorestrictive materials) and composites using these materials have advanced dramatically in recent years. The incorporation of nanotechnologies in the design of such materials is in the concept stage. Compelling military and aerospace applications include ultracompact actuators for navigation of missiles and aircraft, morphing structures such as antenna reflectors that can change shape or reconfigure, robotic structures, and structures that can reestablish their original shape after damage or use.

Important here is the prospect of using materials such as carbon nanotubes as part of the architecture. For example, the extremely high thermal conductivity of carbon nanotubes is cited as a property that could be incorporated into thermal management components on surfaces or embedded in system structures, along with the favorable mechanical properties for lightweight structures.[174] More generally, it is envisioned that nanocomposites specifically tailored for their thermal properties may someday add a useful dimension, thermal management, to complex structures operating in extreme temperature environments.

Another area of multifunctionality that is being pursued at NRL, for example, is the use of structures that also store or provide energy. This includes concepts such as structures with embedded, conformally shaped batteries. Another example is self-consuming structures with solid fuel elements that provide structural stiffness until they are consumed, so that as the stiffness is reduced, so is the mass of the structure.[175] Some of these concepts involve the science of nanoscale materials.

Other connections of multifunction structures to the nanoworld are the ways being explored to improve smart or active materials (such as shape memory alloys and piezoelectrics) by manipulating their properties on the nanoscale or using nanoscale layers and multilayers in the designs. Similarly, schemes to incorporate low-observable, stealthy characteristics to structures are looking to materials science control on the nanoscale to achieve these goals.

A different application area associated with multifunctional structures is self-healing structures. One such concept is the use of self-assembled monolayers and multilayers of organic materials that exhibit inherent self-healing from scratches and similar damage as protective coatings, corrosion inhibitors, or adhesion promoters.[176]

The AFRL is investigating the self-passivating and self-healing properties of polymeric nanocomposites used in connection with the well-known practice of using organic or inorganic fillers to reinforce polymers. Examples of nanoelements used for this purpose are layered silicates and carbon nanotubes. These can be dispersed in resins and processed to produce materials with tailored glassy and/or rubbery moduli. For applications such as thermal protection, polymer-layered silicate materials act as ceramic precursors that form protective ceramic coatings when exposed to high temperatures, effectively self-healing and self-protecting the underlying structures in elevated-temperature environments.

At the University of Illinois, work is reported on the use of composites embedded with fluid-filled particles that break when the structure is damaged and release a reactive polymer that is catalyzed by dispersed catalytic particles within the matrix resin. A substantial fraction of the fracture toughness of woven composites can be recovered in structures that have suffered microcracking using this self-healing method.

In general, multifunctionality in structures will be further enabled by the miniaturization of sensors, electronics, and energy storage and generation devices—a major theme of this report. Owing to their smallness these devices can be more readily embedded and integrated into the framework and structures of systems.

Materials for MEMS

The fact that the mechanical properties of structures of microdimensions can be very different from those of analogous macrostructures has been recognized for some time, but the real practical implications are just now being explored comprehensively. For example, the Materials Research Society has devoted a symposium to the topic of materials science for MEMS every year for 3 or 4 years. It stands to reason that if the mechanical structures themselves are on the order of microns, then the grain structure of mechanical materials is very important and must be understood and controlled precisely at the submicron scale in order to optimize the MEMS mechanical function.

The magnitude and nature of built-in strain in the MEMS structures is extremely important to their practical applications. The grain size and deposition and annealing history are among the important factors in determining residual strain and the effect that this will have on the overall MEMS process. To date, MEMS mechanical properties have been studied most thoroughly in polysilicon[177] and in the sacrificial layers most often used in the MEMS process flows, such as silicon dioxide and silicon oxynitride films.[178] Also being investigated are the effects of metal fatigue and anelastic creep on metal MEMS mechanical structures produced by various processes and involving such materials as Au, Ni, and Ni alloys. Just as for macrosize metal structures discussed previously, the nanostructure of metallic materials resulting from the deposition or forming process used will clearly affect the mechanical properties of MEMS devices. However, in the MEMS size world, these effects will dominate.

Also very important at the micron scale (but not usually on the macroscale) are the effects of surface roughness and surface morphology on the mechanical integrity and strength of MEMS devices. The surface character of the top, bottom, and sidewalls of MEMS structures plays a profound role in mechanical quality factors and other mechanical figures of merit. The control and optimization of all the surfaces of relatively complex mechanical structures throughout the fabrication and packaging processes have been among the most challenging of the technical obstacles to MEMS developers to date.[179]

Many types of materials have been investigated for MEMS applications, including variously doped polysilicon, amorphous Si, single-crystal Si, SiC and other Si alloys, III-V semiconductors, quartz, diamond, and various types of metals and metal alloys. In a number of applications, the limited durability of the MEMS mechanical structure is a severe issue for practical applications. Finding harder, more durable materials than the pervasive silicon and polysilicon is one avenue that is being pursued. Another is to find ways to apply hard coatings to MEMS structures that permit them to withstand surface sliding and mechanical contact.[180] The technical issues involved in hard coatings for MEMS are analogous to those involved in depositing hard coatings on macroscopic structures, but are even more intertwined with material science at the nanoscale.

It is well known that on the microscale, where capillary forces can dominate, surface effects are dramatically different from effects on the macroscale. For this reason, the MEMS developers wage a constant struggle with stiction, the sticking of MEMS surfaces that are meant to be mechanically free, during fabrication, processing, or use of MEMS devices. This phenomenon is significant only at MEMS scales and is a major constraint on yield and operability. In recent years, there has been considerable work toward understanding the forces that cause stiction and toward devising approaches to reduce or eliminate it.[181] Methods for avoiding liquid phase interactions during fabrication such as supercritical drying and dry etching have been developed, and commercial fabrication equipment makers have followed with tools to use in manufacturing of MEMS devices. Hydrophobic coatings have been employed and various surfactant treatments employed to avoid mechanical collapse of the MEMS structures and their pinning to surfaces. This has been a critical technical obstacle to overcome in the application of MEMS and is an example of why the road to mass-manufacturable, high-performance MEMS components has been so long.

Technical Issues and Areas for Development

Although for structural applications it is the mechanical and thermal properties evoked by nanotechnology rather than the electronic and photonic properties used for information and sensor applications, many of the technical issues are similar. The primary areas of technical development supporting improved structural materials include these:

- modeling and simulation
- deposition, fabrication, consolidation or assembly processes for forming the materials
- characterization, testing, and analysis at the nanoscale
- scale-up and manufacturing issues

Some very specific technical issues relating to nanocrystalline materials include the following:

- There is a lack of understanding of the controlling mechanisms as grain size gets below about 50 nanometers for most materials, including thin films and coatings of various types. Viable models, reliable simulations, and quantitative experimental guidelines will have to be developed if materials by design are to be realized for Air Force systems. In addition, effective methods for measuring, classifying, and sorting nanoparticles will have to be developed for use in practical manufacturing processes.
- The instability of the nanosize grain structure at elevated temperatures[182] is the limiting issue in synthesis and manufacturing processes as well as

in structure fabrication and implementation strategies. Grain size growth or subgrain growth may occur at higher processing temperatures or during use at higher temperatures. Clearly, nanoscale-structured materials can be useful only if the nanosize can be maintained.
- The inherent reactivity of very tiny nanoparticles (such as carbon- or aluminum-containing particles) makes producing, handling, and using such materials difficult and potentially explosive. Although this reactivity is the basis for developing nano-scale energetic materials for propulsion or explosives, in the realm of applications for structural materials it is a serious manufacturing problem. In addition to the pyrophoric nature of metallic nanoparticles, in large-scale manufacturing settings there may be health and safety issues associated with handling nanopowder raw materials. Issues like these would have to be addressed when developing manufacturing strategies and implementations.
- The characterization and experimental analysis methods for dealing with nanosize particles, wires, tubes, and ribbons require new tools and new approaches. Measuring the Young's modulus of such a tiny structure as a nanotube would clearly be unlike a conventional measurement. Tools to "see" and manipulate structures on the nanoscale are essential, but their development is still at early stages. The quality control and process monitoring tools that will be needed for full-scale manufacturing using nanomaterials are still further away from realization at this time.
- The myriad of structural materials issues involved in the design, process flow, and assembly of MEMS devices are the key obstacles to deployment of high-performance MEMS components in military systems.

Findings and Recommendations

Finding 3-5. *The application of nanoscience to structural materials is a promising area with important implications for Air Force systems. Such materials could be used for lightweight structures, improved coatings, multifunctional structures, and micromachined structures. Military-specific structural materials applications need special attention by the Air Force.* Many of the emerging nanoscience developments will march ahead without regard to potential Air Force needs or imperatives, and some will find their way into commercial products. However, some military structural materials needs will never be addressed by commercial industry. These include stealthy structures, thermal management structures for spacecraft, lighter-weight military aircraft, and structures for reduced logistics and maintenance costs. However, if the Air Force is to exploit and implement the most compelling advances, even those from the commercial world, then aerospace system designers and manufacturers and their suppliers as well as the Air Force users must constantly be aware of those advances. There need to be mechanisms to link technical developments to military system applications throughout the entire supply chain.

Recommendation 3-5. *The AFRL should continue its strong efforts in structural materials to capitalize on research and development advances. The Air Force should invest in establishing capability on the part of contractors to supply selected military-specific products at the same time as it invests at AFRL to encourage collaboration with cutting-edge researchers and the comprehensive tracking, where possible, of research and development and steering it toward Air Force needs.*

Finding 3-6. *Nanoscience as applied to the structural materials used for MEMS components is key to the successful deployment of MEMS technology.* Unresolved issues such as stiction prevention, sidewall morphology, and durability and stability of micromechanical structures are obstacles to the deployment of reliable MEMS sensors and actuators in military systems.

Recommendation 3-6. *The Air Force should focus development resources on materials issues that currently limit MEMS deployment for the military. These include structural stability, surface durability, manufacturable fabrication processes, and packaging.*

AERODYNAMICS, PROPULSION, AND POWER

Warfare has always required the transport of troops and materials to and across the battlefield. Modern warfare adds the transport of information and energy and extends the range of the battlefield across the globe and out to geosynchronous orbit. Although transatmospheric transport using ballistic missiles and rockets can provide the fastest transfer of personnel and materials across continental distances, air-breathing aerodynamic vehicles will continue to provide the lion's share of rapid transport for the foreseeable future.

Multiple challenges exist for maintaining air and space superiority over the next 50 years. The major challenges will be these:

- to maintain air vehicle survivability—for instance, by means of better IR sensors and flight control—against ground and air-launched threats that incorporate ever-increasing technological sophistication
- to maintain cost-effective, long-range rapid transport capability with increasing fuel prices
- to reverse the trend of significantly increased cost per vehicle for each new generation
- to provide less-expensive, faster-response launch systems

- to protect space-based assets from man-made threats—for example, antisatellite weapons and space debris

Propulsion is at the heart of aerospace vehicle design since the propulsion system and its fuel account for between 40 and 95 percent of the initial system mass. Propulsion is needed to power vehicles ranging in size from grams (for unobtrusive airborne and spaceborne sensors) to hundreds of tons (for large aircraft and launch vehicles). Modern warfare also requires energy—energy in modest quantities for C^3I (watts to kilowatts) and in very much larger quantities for weapons (megawatts and above). Micro- and nanotechnology applied to propulsion and power may offer the opportunity for significant evolutionary improvements to current systems and may enable revolutionary new systems and capabilities.

The technical challenges may be met by utilizing a combination of micro- and nanoelectronics for communications and information processing, MEMS and NEMS for sensors and distributed actuators, and molecularly engineered materials for ultralightweight structures. Batch fabrication, self-assembly, and high levels of functional integration during the fabrication process will be required to make these highly sophisticated systems and subsystems affordable.

Flight Vehicle Aerodynamics

Micro- and nanotechnologies offer the possibility of dramatically improving the aerodynamics of flight vehicles in two principal ways. The first is through direction modification of the aerodynamics by microdevices and the second is through materials changes that achieve more favorable fluid mechanics behavior.

Although the lift and drag of large aircraft is tens or hundreds of tons, the aerodynamics within a few millimeters of the vehicle surface has a profound impact on vehicle performance. This so-called boundary layer is the result of the viscous nature of air and the high characteristic speeds and length scales of man-made vehicles. It has long been known that local manipulation of the boundary layer can radically alter first-order vehicle performance parameters such as drag and stability. A relatively recent research field is the active control of fluid mechanics, which uses high-frequency, dynamical control of aerodynamic surfaces to alter the mean aerodynamic behavior. Subscale experiments have demonstrated that such active control can be used to alter the turbulent nature of the flow, perhaps leading to laminar flow vehicles with dramatically increased range and payload compared with existing aircraft. A second flow control application uses unsteady manipulation of the boundary layers to generate large forces and moments to achieve control and maneuverability of the flight vehicle. These controllers would supplement or replace conventional large-scale control surfaces, their actuators, and hydraulic lines and valves. This would alter the steady-state performance versus controllability design trades, bringing advantages such as reduced drag, weight, and observability.

Most approaches to active fluid control require large numbers of sensors and actuators distributed over the aerodynamic surfaces to locally influence the flow. The sensing and actuating frequencies required are generally on the order of 1,000 Hz. For this reason, MEMS approaches appear very attractive since they could allow low-cost manufacturing of large numbers of thin-film systems containing sensors, actuators, and local computation. Also, while the sensing requirements may be met with multiple technology approaches, the high actuation frequencies require very small, low-mass machines—in a word, MEMS.

MEMS-based control of aerodynamic surfaces has been demonstrated by researchers at the University of California at Los Angeles and at the California Institute of Technology under a DARPA-sponsored program. In the case of a delta wing under flight, the naturally occurring pair of primary vortices provides a mechanism for amplified distributed control. These vortices start near the leading edge and convect down the wing. The genesis location of these vortices determines the overall characteristics of the resulting large primary flow structures, and all six components of forces about an aircraft can be controlled.[183] Wind tunnel experiments using a 56.5 degree sweepback delta wing with a 31.8-centimeter root chord have shown that distributed 2-millimeter MEMS "bubble" actuators on the leading edge can provide positive roll control comparable to that obtained using conventional trailing edge flaps that are two orders of magnitude larger.[184] MEMS fabrication techniques also enabled the development of distributed shear stress sensor arrays for high-speed measurement of two-dimensional shear stress patterns on wing surfaces. Current research is focused on determining control laws for unsteady flight control. The goal is to make a delta-wing aircraft maneuver without using a rudder or any other macroscopic control surfaces.

Technical issues for MEMS-based aerodynamic control include scalability to full-size flight vehicles, suitability for supersonic and transonic vehicles, and robustness of flight control under real-world dust and icing conditions. These issues must be addressed by full-scale flight-testing programs.

Vehicle Health Monitoring

As mentioned in a previous section on sensors, distributed MEMS-based sensors can be used to measure pressure, velocity, and shear stress distributions on aerodynamic surfaces to aid in flight vehicle development and qualification. In the longer term, similar sensor arrays could be permanently integrated into flight hardware to provide continuous health and status monitoring. Distributed stress sensors could provide stress cycle information that would be used to determine when and if airframe components need to be replaced. During battle, distributed stress and vibration sensors would provide real-time assessment of impact damage and distribution. Note that this technology would be applicable to both air vehicles and launch systems and could be readily integrated with MEMS-based aerodynamic control. Technical issues include the robustness of individual sensors to real-world flight conditions (the Boeing pressure sensors had to be redesigned to withstand particle impacts), the real cost of add-on or integrated sensor networks, and definition of a standard sensor network protocol for future air vehicles.

Air-Breathing Vehicle Propulsion and Power

Aerospace dominance is one mission of the Air Force—that is, to command the sky from the troposphere to interplanetary space. Current air-breathing, turbomachinery-based propulsion systems operate mostly in the troposphere and occasionally in the stratosphere. The promise of micro- and nanotechnology can be considered in terms of evolutionary improvements to current gas turbine approaches and as an enabler of new, very small propulsion and power systems. These very small propulsion and power systems could be used singly in very small aerospace vehicles or arrayed to provide scalable thrust and power for larger vehicles.

Modern gas turbine engines require parts and subsystems that must operate reliably for thousands or tens of thousands of hours in highly erosive, oxidizing, very-high-temperature conditions, over 2000 K in the main gas path and above 700 K even in the nacelle outside the engine. This extraordinarily harsh environment severely restricts the range of materials and designs that can be considered. Thus, most engine materials have been specifically developed with propulsion system applications in mind. Development of micro- and nanomaterials is likely to follow a similar path.

Evolutionary Improvements

One class of evolutionary improvements to gas turbines is MEMS-based flow control concepts, which are in many ways similar to those discussed for air vehicle aerodynamics. Improvements to the steady-state aerodynamics through flow control of gas turbine propulsion systems could include inlet and compressor end wall aerodynamics. Several projects are currently investigating unsteady control of inlet and nozzle flows. Should they be successful, these approaches would offer such benefits as reduced weight, signature, and fuel consumption, with range–payload system improvements of 5 to 10 percent. Another concept uses flow control actuators mounted above the compressor and turbine rotor tips to dynamically synthesize a flowfield resembling that of a much tighter tip clearance, offering significant efficiency and life improvements. Such flow control is more difficult in the turbine due to the much harsher environment.

A somewhat different application of MEMS technology offers new approaches to old gas turbine problems, namely the dynamic system instabilities that plague gas turbines, including rotating stall and surge, combustor instabilities, airfoil flutter, and inlet instabilities. The control or suppression of these instabilities offers improved range payload (by 10-15 percent for surge and stall), reduced emissions, and reduced maintenance. Although the underlying physics and implementation details may vary, the control implementations for these instabilities share the common elements of distributed sensing and actuation. While there are no specific requirements for a MEMS-scale device, the actuation frequencies needed are difficult to achieve with other technologies. Also, technology solutions using large numbers of sensors and actuators are usually quite sensitive to unit cost, which could be low with micro- and nano-approaches.

Micro- and nanotechnologies may also have much to contribute in the area of engine controls and accessories. The control system of a modern aircraft engine now accounts for over 15 percent of the acquisition cost and 40 percent of the maintenance and overhaul costs. One major engineering challenge for these systems is the high temperature, which currently requires that many of the sensors and electronics be located remotely in a cooled environment, thus increasing complexity and cost and discouraging redundancy. High-temperature microsensors and accompanying electronics would be very attractive from a systems viewpoint and might be an enabling technology for some of the flow control schemes discussed above. SiC- and GaN-based, high-temperature micro-electronics and MEMS is the most common research direction aimed at solving this problem. Another approach might use lower temperature electronics locally packaged with chip-level MEMS-based coolers.

Gas turbine fuel controls consist of two principal elements: a computer and a fuel-metering unit controlled by the computer, which consists of redundant transducers, valves, and actuators. These units are now custom engineered for each application, so that many different models are manufactured in small quantities and at great cost. MEMS offers the possibility of developing a standard (albeit very small capacity) integrated fuel management unit, numbers of which operating in parallel would provide the fuel flow required for different size engines. This approach could offer the benefits of lower cost and higher reliability (since the multiple units operate in parallel). Fuel pumps might also be amenable to a similar approach (at least for small engines).

Technical issues for these evolutionary improvements to air-breathing turbomachinery include the integration of MEMS sensors and actuators onto curved surfaces, the transmission of data across rotating surfaces, the development of instability control algorithms and active control for specific applications, and the ability of MEMS sensors and actuators to survive high-temperature chemically reacting flows without fouling. Testing and diagnostic efforts using relevant turbomachinery need to be initiated.

Revolutionary Opportunities

Micro- and nanotechnologies may enable air vehicle propulsion and power systems orders of magnitude smaller than today's engines. A MEMS approach to propulsion systems on the millimeter or centimeter scale offers several tantalizing advantages. First, the power-to-weight ratio can improve dramatically as size is reduced since engine power scales with the intake area while mass scales with volume. Thus, all else being equal, power-to-weight increases linearly as size is reduced. In other words, this scaling maintains the propulsion system diameter per unit thrust (since the same air must be ingested per unit thrust) but shrinks the length to a few millimeters. Also, the small size facilitates the use of materials with high-temperature properties similar to those of conventional superalloys, such as single-crystal silicon and silicon carbide. Second, the adoption of semiconductor industry wafer-level micromachining approaches means that complex, high-precision parts and assemblies can be realized at very low unit costs.

One example now in laboratory development is a 4-millimeter-thick, 1-cubic-centimeter MEMS-based gas turbine engine, illustrated in Figure 3-14. Figure 3-15 shows a scanning electron micrograph of the batch-fabricated silicon turbine. Initial units are designed to produce about 100 millinewtons of thrust; later units may provide up to 10 times more in similar packages (this is equivalent to 10-100 watts of shaft power). The first units are expected to have quite poor fuel efficiency, on a par with early jet engines, but quite high thrust-to-weight ratios. At our current level of understanding, it seems unlikely that MEMS power systems will achieve fuel consumption equivalent or superior to the best current large systems, but power-to-weight or thrust-to-weight ratios 10-100 times that of the best large systems may indeed be achievable.

Many applications for these microscale propulsion systems can be envisioned. One of the first might be the micro air vehicles (MAVs) under development by DARPA for surveillance and reconnaissance applications. AeroVironment, Inc., has flown the 6-inch wingspan Black Widow MAV for 30 minutes using an electric motor powered by batteries.[185] It has a loiter velocity of 25 mph

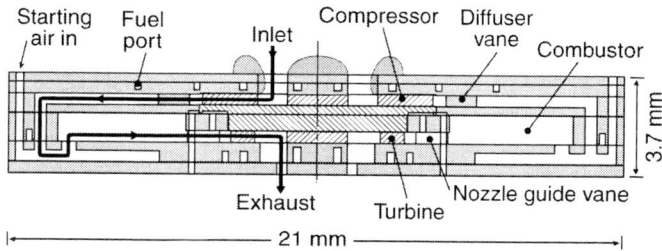

FIGURE 3-14 Micromachined gas turbine engine. Courtesy Massachusetts Institute of Technology.

FIGURE 3-15 Silicon turbine from the micromachined gas turbine engine. SOURCE: Epstein, A.H., S.D. Senturia, O. Al-Midani, G. Anathasuresh, A. Ayon, K. Breuer, K-S. Chen, F.F. Ehrich, E. Esteve, L. Frechette, G. Gauba, R. Ghodssi, C. Groshenry, S.A. Jacobson, J.L. Kerrebrock, J.H. Lang, C-C. Lin, A. London, J. Lopata, A. Mehra, J.O. Mur Miranda, S. Nagle, D.J. Orr, E. Piekos, M.A. Schmidt, G. Shirley, S.M. Spearing, C.S. Tan, Y-S. Tzeng, and I.A. Waitz. 1997. Micro-heat engines, gas turbines, and rocket engines—The MIT microengine project, AIAA Paper 1997-1773. Reston, Va.: American Institute of Aeronautics and Astronautics.

and a total mass of 57 grams, and it transmits live color images to a ground-based operator. An order-of-magnitude increase in cruising time is required, indicating the need for micro-combustion engines or generators. A single 100-millinewton thrust engine is sufficient to power a 50-gram gross takeoff weight vehicle (about 6-inch characteristic length) with a 50-kilometer range at 60 knots flight speed. More advanced MAVs could use five to ten engines to generate vertical lift directly, thus enabling both 60-knot standard flight using efficient wing-generated lift plus hovering and low-speed capability for maneuvering within buildings and caves. Insect-sized flyers, an order of magnitude or more smaller than the MAVs currently under development may be feasible within the next two decades.

Batch-fabricated micro-turbine engines can be used on larger vehicles as well since multiple engines can be used as vehicle size increases. Several thousand engines would be needed for a cruise-missile-size vehicle. The high power-density microengine approach is particularly attractive for volume-limited applications such as air-launched munitions. At this scale, thrust would become a commodity in that a common engine design would be used in variable numbers as demanded by the particular vehicle design. This is in contrast to the current practice, which requires a custom engine design for each application with relatively small numbers made of each engine model. The new approach could facilitate more rapid vehicle development and might dramatically reduce propulsion system costs. Longer-term applications would entail the use of very large numbers of microengines (tens of thousands or more) as lift engines for larger aerospace vehicles such as unmanned combat air vehicles (UCAVs). Advanced engine arrays may develop thrust densities of 100 pounds per square foot, levels comparable to the steady-state aerodynamic lift on the wings. This approach offers the advantages of extraordinary redundancy and reduced weight at the cost of installation, interconnection, and control challenges.

The use of MEMS for aerodynamic control, propulsion, and power requires new technologies for packaging and installation. Some of those needed on an individual chip level are now being pursued in conjunction with development of the microdevices. Additional integration issues arise if such chips are to be used in large numbers. The design for installation of large chip arrays is just being considered by airframe and system designers. Such studies will most likely identify new R&D requirements in this area. For example, DARPA has recently held workshops on the future of MEMS, in which research on "MEMS by the yard" (i.e., low-cost manufacture of large areas) was called for.

Air Vehicle Electrical Power and Auxiliary Systems

In addition to improving the propulsion system, advances in micro- and nanotechnologies may also dramatically alter the nature of other air vehicle subsystems. As an example, boundary-layer flow control as an effectuator of vehicle control (discussed above) could obviate the need for aircraft hydraulics, a subsystem that is very complex and maintenance-intensive, becoming a nontraditional enabling technology for an all-electric aircraft.

Large air vehicles use shaft power takeoff from the propulsion system to generate electric power for onboard needs. If sufficiently efficient, microscale power generators such as the gas microturbine engine generator discussed above could locally provide the needed power, thus reducing weight, simplifying the aircraft electrical system, and adding redundancy. Key electronic systems such as receivers, high-accuracy clocks, and GPS receivers could remain active even in parked aircraft; this would eliminate warm-up periods and resynchronization of onboard systems. For missiles and munitions, the fuelled micropower generators would replace batteries (hydrocarbon fuels have 20 times the energy density of the best battery chemistry) with significant weight savings. DARPA has a number of ongoing micropower (milliwatt-to-watt) development projects that include (1) gasdynamic (microturbines, micro Wankel engines, etc.) and thermoelectric energy conversion that use hot gas from microcombustors, and (2) hydrocarbon fuel reformers that would enable fuel cells to operate using normal liquid fuels. Participants in the micro power generator (MPG) program include the Massachusetts Institute of Technology, California Institute of Technology, Princeton, University of California at Berkeley, Case Western Reserve University, Georgia Institute of Technology, the University of Southern California, Pacific Northwest National Laboratory, Honeywell, and the Aerospace Corporation.

High-efficiency MEMS-based refrigerators, packaged within microelectronics chips, may reduce the scope of, or even eliminate the need for, the complex avionics cooling systems now used on high-performance military aircraft. This technology would also reduce the weight and cost and improve the operability of sensor coolers on aircraft, ordnance, and spacecraft.

Launch Vehicle Propulsion

Solid Propellant Rocket Motors

Solid rockets use a pressure vessel that surrounds a combusting solid to direct the emerging hot gas through a converging/diverging nozzle. Each pound of mass that can be removed from the pressure vessel, or casing, allows an extra pound of payload to be delivered. Metal casings have given way to composite casings made of carbon or glass fibers, which have higher strength-to-weight ratios. Since solid rockets burn for a limited time once ignited, a layer of internal insulation is used to shield the casing from high-temperature exhaust; this allows composites with polymer binders to be used. An obvious application of micro- and nanoengineering to solid rockets might be the use of carbon nanotube composites if significantly improved strength-to-weight ratios are achievable; this could significantly decrease casing mass. The technical challenge is to fabricate carbon nanotubes of suitable length (millimeters or even centimeters) in large enough quantities at reasonable cost (see the preceding section).

The propellant in solid rockets is a mechanical mixture of oxidizer, fuel, and binder solids. When the local temperature gets sufficiently high, an exothermic chemical reaction takes place. The key is to create a sufficient activation barrier between highly energetic components so that both species can coexist at typical storage and handling temperatures. Micro- and nanoengineered coatings may enable more energetic species with overall higher performance to coexist under normal conditions.

High-performance solid rocket propellants usually include aluminum powder as fuel. Typical large solid rocket engines contain 14 to 18 percent aluminum powder by weight. Nanopowder aluminum provides more rapid combustion owing to its increased surface-area-to-volume ratio, which results in faster linear burn rates. Burn rates for operational rocket propellants are typically between 0.3 and 0.8 inches per second, which would generate unacceptably long burn times (~40 minutes for the 110-foot-long solid rocket motor units (SRMUs) used on the Titan IV booster if the propellant burned from bottom-to-top—like, for instance, a cigarette). Large, solid rocket engines have propellant cores cast with convoluted openings to provide radial burning with larger surface areas; this leads to higher gas generation rates and shorter burn times (about 2 minutes for the Titan IV solid boosters). Nanopowder propellants offer much higher burn rates, so simplified core designs with higher average propellant density could be used. Technical issues include determination of nanopowder oxidation rates in contact with the fuel/oxidizer to establish propellant storage lifetimes, control of engine instabilities with faster-burning propellants, and the possible use of coatings on the nanopowders to chemically stabilize them.

Liquid Propellant Rocket Motors

Liquid propellant rocket engines offer higher exhaust velocities than solid rockets in return for increased complexity and cost. Higher specific impulse translates into reduced propellant mass requirements for any particular mission. Liquid hydrogen and oxygen provide the highest practical specific impulse (about 450 seconds), while liquid hydrazine (or its derivatives) and nitrogen tetroxide offer noncryogenic storage with hypergolic ignition for increased reliability at reduced specific impulse (about 320 seconds). Payload delivery performance is a function of specific impulse, wet/dry mass fractions, and the number of stages used. Micro- and nanoengineered materials with increased strength-to-weight ratios, when realized, will improve the wet/dry mass fractions and result in more delivered payload per unit launch weight.

A complete liquid propulsion system consists of a thrust chamber and nozzle, propellant piping and controls, and pumps or a high-pressure gas system (to pressurize the propellant tanks), which feeds propellants to the combustion chamber. With current MEMS technology it is now feasible to micromachine all of these components at the millimeter scale. The first MEMS dime-sized, 3-pound-thrust bipropellant rocket engine has been tested. This device, developed at the Massachusetts Institute of Technology with NASA funding, was fabricated out of six bonded, micromachined silicon layers; it uses oxygen and methane as propellants. Owing to its small size and favorable scaling relationships, its thrust-to-weight ratio is far greater than that for traditional bipropellant thrusters. Future engines may be developed at larger or smaller thrusts as the system requirements and technology allow. Another example of a smaller, less ambitious bipropellant thruster is the "microjet" developed in the United Kingdom at the Defence Establishment Research Agency. This 13-millimeter-long, 2-gram-mass thruster produces 63 millinewtons of thrust using hydrogen peroxide and kerosene. Obviously, multiple units can be used in parallel to achieve higher thrusts.

A revolutionary application enabled by MEMS liquid rocket engines would be microlaunch vehicles and ballistic, multistage missiles in the size range between 15 and 800 kilogram gross liftoff weight (GLOW). Design studies suggest, for example, that a 60-kilogram GLOW two-stage rocket could deliver a kilogram or two to low Earth orbit or twice that as a ballistic payload to a 4,500-nautical-mile range. Except for the seeker, these vehicles would be about the size and complexity of a tactical missile and thus should cost about the same. They could be ground or air launched. As a launch vehicle, the microrocket redefines the concept of low-cost access to space as cost per mission rather than cost per pound of payload (these are about the same for large rockets). Thus, it will be possible to place a payload (albeit a small one) into orbit for $10,000 to $50,000 rather than the $10 million to $50 million cost today. This cost is low enough that launchers can be stockpiled for on-demand launch access to space and deployed across the planet for tactical missions. Aggressive orbital applications might include such military missions as visual and IR inspection of space objects, jamming of communications satellites, electronic intelligence gathering, and anti-satellite operations. More peaceful military applications could include distributed space weather monitoring during geomagnetic storms and low-resolution (>50-meter ground resolution) Earth observation. Presumably, scientific and commercial users of space would adapt their missions to take advantage of this very low cost as well. Suborbital uses might include ultralong-range micro tactical ballistic missiles with sensor or nonnuclear munitions payloads, ballistic imaging or electronic intelligence payloads that loiter for 10 minutes above a battlefield, and very small antiballistic missile interceptor missiles. Micro- and nanoelectronics and MEMS can provide ultrasmall intelligent payloads that are ideally suited to micro launch vehicles.

Spacecraft Propulsion

Chemical Propulsion

MEMS-scale liquid propellant rocket engines offer significant evolutionary advantages over conventional engines and also enable new space systems concepts. MEMS-based thrusters with ultrasmall micro- to millinewton thrust levels are needed for emerging micro-, nano-, and picosatellites. Thrusts of 1 to 5 pounds (4.5 to 23 newtons) are currently used for station-keeping by large geosynchronous satellites, while hundreds of pounds of thrust are needed for impulsive orbital maneuvers such as the apogee kick burn required to transfer a satellite from geosynchronous transfer orbit (GTO) to final GEO. Thrusters for kinetic-kill vehicles can fall in between. Cube-square scaling implies that a MEMS machine will have at least an order of magnitude higher thrust-to-weight ratio than a large rocket engine. The reduced weight can be used for increased payload or for redundancy to improve reliability. MEMS also facilitates the commoditi-

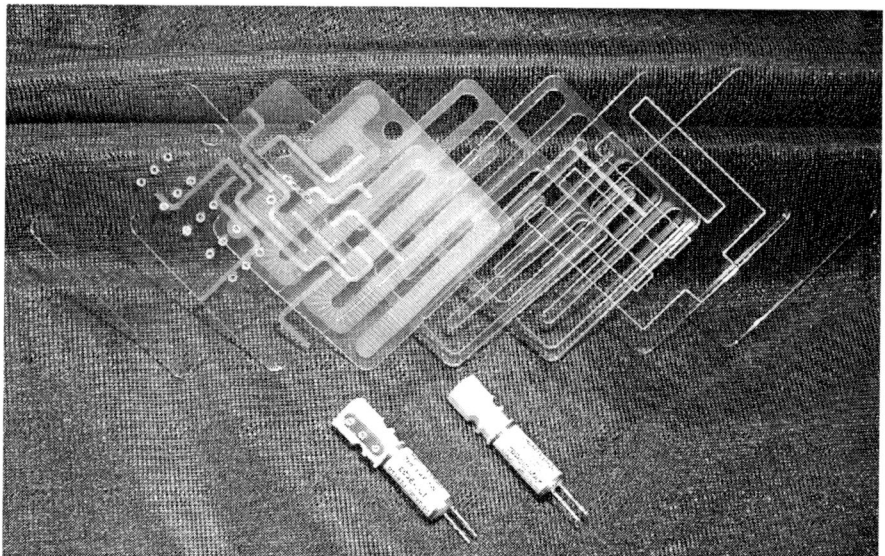

FIGURE 3-16 Planar glass layers for a batch-producible cold gas propulsion module. SOURCE: Huang, A., W.W. Hansen, S.W. Janson, and H. Helvajian. 2002. Development of a 100-gm-class inspector satellite using photostructurable glass/ceramic materials. Pp. 297–304 in Photon Processing in Microelectronics and Photonics, Proceedings of SPIE Volume 4637. K. Sugioka, M.C. Gower, R.F. Haglund, Jr., A. Pique, F. Traeger, J.J. Dubowski, and W. Hoving, eds. Bellingham, Wash.: The International Society for Optical Engineering.

zation of thrust. Significant system advantage can accrue if a standard microrocket module—well developed, well characterized, and manufactured in quantity inexpensively—is adopted for multiple applications, with different numbers of engine modules utilized, depending on the mission. This has the potential to dramatically reduce the cost of space propulsion. Also, the very small size of individual motors yields very fast start-up and shutdown times, which allow a highly precise impulse increment to be imparted to a vehicle.

An example of prototype thruster system components that could be produced by batch-fabrication techniques is shown in Figure 3-16. Six micromachined glass layers form a liquid storage tank, gas/liquid separator, gas plenum, gas distribution plumbing, and nozzles for a cold gas propulsion system. While this set of glass wafers was fabricated using direct-write laser-patterning of Foturan™ glass, mass production would utilize Foturan™, planar masks, and UV exposure much like the photopatterning step in the fabrication of semiconductor wafers. The layers could also be fabricated out of silicon using photolithography and deep reactive ion etching. This stack utilizes five miniature solenoid valves with six nozzles to provide translational thrust along two axes and rotation about the third axis. MEMS valves would be preferred for batch-fabrication capability, but miniature solenoid valves currently provide lower-power operation with lower leak rates than MEMS valves. The 10-millinewton thrusters are appropriate for translational and rotational control of a ~1 kilogram mass co-orbital satellite assistant.

A revolutionary thruster approach enabled by MEMS is digital propulsion, which consists of an array of single-shot thrusters that individually produce only one impulse each; spacecraft maneuvers are performed by firing unused thrusters at specific locations at the right times. Microfabrication enables the creation of large arrays of addressable thrusters, e.g., 10,000 on a 10-centimeter-square surface using 1-millimeter center-to-center spacing. This digital thruster system is planar, is scalable in area, does not require separate propellant tanks or plumbing, does not require microvalves, and can function as a structure. Solid rocket, water electrolysis, and subliming solid thruster arrays consisting of micromachined glass and silicon layers have been fabricated and tested under a DARPA program.[186] Measured impulse bits are 0.1 millinewtons for the solid rockets. A 3 × 5 thruster array flew on the space shuttle Columbia during the STS-93 mission. Similar digital thruster efforts have been funded by AFOSR and the French national space agency.[187,188]

Electric Propulsion

Electric thrusters use electric power to accelerate propellant molecules, atoms, or ions. The advantage of this approach is that the specific impulse can be more than an order of magnitude greater (600 to 5,000 seconds) than that of chemical thrusters (50 to 450 seconds). This translates into significantly reduced propellant requirements for a given mission if sufficient electric power is available. Power requirements can range from tens of milliwatts for attitude control of a 1-kilogram-mass spacecraft to tens of kilowatts for orbital maneuvering of ~3000-kilogram-class spacecraft. Power requirements scale with thrust and the square of specific impulse.

Figure 3-17 shows a plot of spacecraft power levels versus first year of launch for different series of Intelsat communications satellites over the last 20 years. The spin-stabilized VI series was not included since its configuration was different from that of the other three-axis stabilized series. Note that the trend is toward increased power with time and that the trend appears to be accelerating. Commercial spacecraft with 11-kilowatt power levels, e.g., the Thuraya communications satellite,[189] are now routine, and spacecraft with 15- to 20-kilowatt power levels are being designed. Note that spacecraft power levels of 1 to 10 kilowatts are becoming available for electric thrusters. Air Force communications satellites, e.g., the flagship MILSTAR, with ~5 kilowatts of solar power, typically operate at lower power levels and lower power-to-weight ratios than commercial communications spacecraft. Nevertheless, they follow similar trends

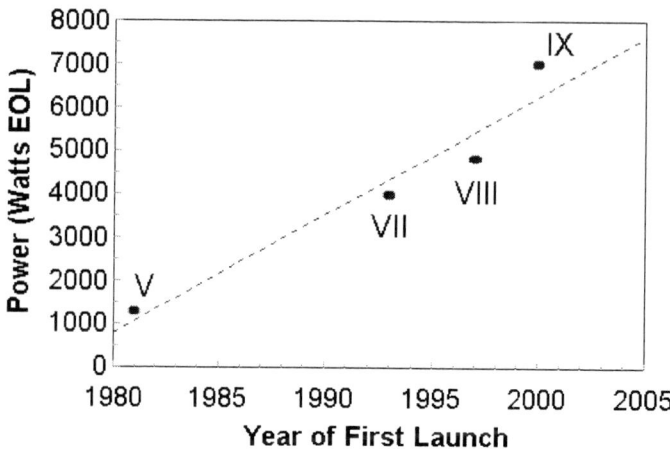

FIGURE 3-17 Spacecraft power for INTELSAT satellites. Data from Martin, D.H. 2000. Communications Satellites, 4th ed. Reston, Va.: American Institute of Aeronautics and Astronautics, Inc.

and are expected to more closely match their commercial counterparts as mission needs, e.g., radar and direct broadcast of video data, converge.

Micro- and nanofabrication enables a radically new form of high-specific-impulse electric propulsion that uses field ionization or field evaporation to ionize propellant atoms or molecules, which are subsequently accelerated by an applied electric field to generate thrust. Field ionization and emission are electron tunneling phenomena that require electric fields of 10^7 to 10^8 watts per centimeter, which can be generated using 100 to 1,000 volts across a 0.1-micron gap. If the emitting surface has a radius of curvature of ~30 nanometers or less, local electric field enhancement occurs, dropping the required voltages to the 10- to 100-volt range. Micro- and nanofabrication enables field emission and ionization using tens of volts instead of kilovolts. Conventional ion engine design requires magnetic plasma confinement, multiple power supplies, and roughly a hundred piece parts such as accelerator grids and capacitors, while a field-ionization-based ion engine would require no magnetic field, only two power supplies, and fewer than 10 piece parts. This approach creates an essentially two-dimensional ion engine that is scalable in area and thrust, thus commoditizing high-specific-impulse thrust for spacecraft.

Field emission sources are currently under development as an enabling technology for flat panel displays, simplified microwave tubes, and vacuum microelectronics.[190,191,192] Examples of field emission sources include Spindt cathodes, diamond like carbon coatings, and carbon nanotubes.[193,194,195] Field emission sources are simple and robust because they do not require a heating element or a plasma discharge to function. Field emitters will find use in electric propulsion systems as an electron source for neutralizing ion beams that generate thrust. Microfabrication offers similar structures for the efficient production of ions.

Liquid metal ion sources exploit the instability of conducting liquid surfaces when electric fields of 10^5 to 10^6 volts per centimeter are applied.[196] Electrodynamic forces cause the liquid surface to form one or more Taylor[197] cones that have sharp tips with radii of curvature between 5 and 50 nanometers, thus generating local electric fields in excess of 10^7 volts per centimeter through field enhancement. Johannes Mitterauer proposed miniaturized liquid metal ion sources that could be fabricated using microfabrication technology in the early 1990s.[198,199] These revolutionary devices were based on the Spindt microvolcano, which is basically a reverse-polarity Spindt cathode with a hole drilled through the emitter to feed gas or liquid to the high-field region near the gate.[200] The advantage of this microfabrication approach is that gross electric fields of 10^6 volts per centimeter may be readily generated by applying 100 volts across a 1-micron-wide gap. Since the field evaporation/ionization process is fairly efficient (less than 10 eV loss per ion), ion engines with high thrust efficiency at low specific impulse, e.g., 1,300 seconds for indium, become possible. A planar, scalable electric propulsion system that can used for picosatellites through 5,000-kilogram-mass satellites appears to be possible.

Tether Propulsion

A tether connecting two masses in orbit will always try to line up with the radial gravity vector. This effect, called gravity-gradient stabilization, is used to create passive Earth-pointing stabilization for many microsatellites. Electrodynamic tethers exploit Earth's magnetic field and the local space plasma environment in LEO to generate either power or thrust; they can operate either as motors or electric generators. In the power generation mode, the v × B Lorentz force drives a current in an Earth-pointing conducting tether when the tether velocity has a component perpendicular to the local magnetic field. Electrons are emitted at one end of the tether and collected at the other; the space plasma provides the return current. Orbital kinetic energy is converted into electric power; this can be used to provide emergency power, to provide primary power for short-duration missions, to provide intermittent power to charge laser or microwave weapons, or to provide active deorbit capability. In the thrust generating mode, a current is forced through the tether to generate a j × B force perpendicular to the local magnetic field direction. This is an electric propulsion system with infinite specific impulse.

Tethers can also operate as momentum exchange systems to provide momentum and energy transfer between attached masses. Proposed systems essentially catch a payload at one end and throw it away at higher velocity, as shown in

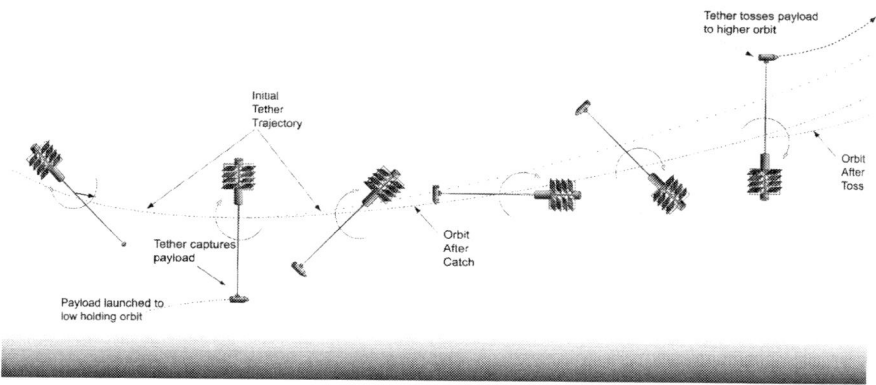

FIGURE 3-18 Use of a momentum-exchange tether to perform an orbit transfer.
SOURCE: Tethers Unlimited, available online at: <http://www.tethers.com/MXTethers.html> [May 22, 2002].

Figure 3-18. The tether system loses energy and drops into a lower orbit, while the payload gains energy. Note that a momentum and electrodynamic tether can be combined into a single entity to enable rapid transfer of a payload into another orbit, followed by slower electrodynamic orbital maneuvering of the tether system to return it to its original orbit.

A number of tether-based momentum-exchange concepts such as the LEO-to-GTO tether boost facility have been studied as possible long-term space transportation systems that would efficiently transfer spacecraft between orbits without using propellant.[201] Applications in the 20- to 50-year time frame include payload capture from hypersonic air-breathing vehicles with subsequent transfer into LEO, transfer of payloads from LEO to a variety of orbits, and transfer from LEO to Earth escape trajectories, e.g., to lunar transfer orbit. A key feature of these systems is the need for tethers with tens to hundreds of kilometer lengths. The LEO-to-GTO tether boost facility mentioned above would require a 100-km-long tether of several millimeters diameter and a mass of 8,300 kilograms. The tether material was Spectra 2000™, a highly oriented polyethelyne with a tensile strength of 4 gigapascals and a density of 0.97 grams per cubic centimeter.[202] A carbon nanotube-based composite could cut this mass by more than an order of magnitude, resulting in lower system launch masses or increased tether lengths for higher energy transfers.

Conductive carbon nanotubes might be ideal for electrodynamic tethers, while semiconductor carbon nanotubes could be ideal for momentum exchange tethers. A carbon nanotube-based structure that could switch conductivity on demand would be even better. Owing to the degrading effects of the space environment, e.g., radiation damage and micrometeor impact, self-healing or repair would also be helpful.

Space Power Generation

Power for avionics, electric actuators, and electric propulsion systems is produced by shaft-driven generators, photovoltaic cells, thermocouple arrays, thermionic arrays, fuel cells, or batteries. Output power levels can range from microwatts to tens of megawatts. Micro- and nanotechnology can be applied to all of these systems to produce miniature (microwatts to watts) power systems and to improve the performance of fuel cells, thermionic converters, and batteries.

The greater part of electric power generation in space is done by solar cells. The principal exceptions have been nuclear thermoelectric generators (mainly for spacecraft operating beyond Jovian orbit) and nuclear reactors on a few large Soviet military satellites. Electrodynamic tethers are a possible challenger for future mission applications that can tolerate, or benefit from, rapidly decreasing orbital energy (altitude). For a 500-kilometer altitude orbit, the change in total energy is about 4.2 kJ/kg-km; dropping 1 kilometer in altitude per hour would release 4,200 joules of energy per kilogram of spacecraft at an average power level of 1.2 watts per kilogram, assuming 100 percent efficiency for the conversion of potential energy to electric energy. Orbit lifetime from this altitude at this energy production rate would be about 2 weeks. Longer missions will clearly benefit from solar cells.

Solar Cells

Solar cell technology currently provides sunlight-to-direct-current conversion efficiencies of 15 percent for silicon cells and 18 percent for gallium-arsenide cells. Photons with energies below the semiconductor bandgap energy are not absorbed, photons with energies just above the bandgap energy are converted to free electrons with near 100 percent efficiency, and photons with even higher energy dump their excess energy as heat. Multiple-junction cells with 40 percent conversion efficiency are possible, but this approach results in a complicated stack of three or more cells with totally different substrates (these provide multiple bandgap energies for efficient conversion at multiple places in the solar spectrum) and mechanical stress-matching layers. Nanoengineered materials may provide better conversion efficiencies by allowing the use of multiple quantum well absorbers.

Solar Dynamic Electric Power Generation

MEMS opens up other possibilities that might be attractive for some applications. Solar dynamic systems had been considered unfeasible below 50 or 100 kilowatts, considerably larger than any current applications except for the International Space Station. In the long term (20 to 50 years), however, 50- to 100-kilowatt communications or radar satellites may be commonplace. For micro-, nano-, and picosatellites, the system mass advantages inherent in MEMS devices such as micro turbogenerators suggest that solar dynamic systems may be feasible down to a few watts in size. Given the high conversion efficiencies of the best solar cells, however, the solar dynamic generation may be most attractive when used in a hybrid system with solar cells.

A different space power application would combine the MEMS bipropellant rocket engine technology with MEMS turbine generators to construct a fueled power unit. Running on station-keeping fuel, a unit such as this might be useful as an emergency power source when there is a problem with solar array deployment, providing sufficient time for ground operations to diagnose and perhaps fix the problem before the onboard batteries run out. A nonmilitary application would be use in planetary missions if nuclear power systems prove politically unattractive.

Batteries

MEMS and microtechnology enable miniaturized combustion systems and turbomachinery, while nanotechnology enables improved performance for individual components, particularly those that rely on electrochemistry. Nanostructured electrodes for fuel cells, batteries, and supercapacitors are an ongoing research area. Decreased anode-cathode gaps enable faster ion and electron transfer with reduced internal impedance and increased power handling capability. Nanostructured electrodes and electrolytes should enable more efficient charge transfer and storage, resulting in improved capacity.

The ubiquitous carbon nanotube has appeared as a possible improved anode material for lithium-ion batteries. Graphite is currently used as the lithium-storage anode; it has a storage capacity of one lithium ion for six carbon atoms. Recent research by Otto Zhou and colleagues at the University of North Carolina has shown that carbon nanotubes apparently have twice the lithium storage capacity per carbon atom. However, the actual performance limits and cost of lithium-ion batteries using carbon nanotubes have yet to be determined.

Rechargeable batteries are essential for spacecraft with multiyear lifetimes. Current space-qualified technologies include nickel-cadmium and nickel-hydrogen batteries with energy storage densities of ~30 and 50 W-hr/kg, respectively. Lithium-ion batteries for terrestrial applications have energy storage densities of ~100 W-hr/kg, but they do not function well at temperatures below $-20°C$. The theoretical energy density of lithium-ion batteries is ~300 W-hr/kg, a sixfold improvement over existing space-qualified batteries. Nanotechnology in the form of carbon nanotube anodes may help to reach this goal, but more research is needed.

Findings and Recommendations

Finding T5. *Emerging microtechnology offers new opportunities in propulsion and aerodynamic control, in particular in (1) distributed sensors and actuators on both macro-aerodynamic surfaces and macro-aeropropulsion units and (2) new, scalable, miniaturized and distributed aero- and space-propulsion systems.* Emerging microtechnology has achieved preliminary success in sensing and controlling the boundary layer on full-size, subsonic airfoils. New devices for controlling gas and liquid flow, fabricated using microtechnology, promise to increase the power and reliability of air-breathing, full-size propulsion units. Several new aeropropulsion and space propulsion systems, such as micro-turbine engines and micro-rocket engines, have been fabricated and are in the early test phase.

Recommendation T5. *The Air Force should move decisively to develop new research and development programs to bring microtechnology to both macro- and microscale propulsion and aerodynamic control systems.*

REFERENCES AND NOTES

1. United States Air Force. 2000. Air Force Strategic Plan, Volume 3: Long-Range Planning Guidance. Washington, D.C.: United States Air Force.
2. Office of the Secretary of Defense. 2000. Space Technology Guide FY 2000–2001. Washington, D.C.: Office of the Secretary of Defense, Assistant Secretary of Defense (Command, Control, Communications, and Intelligence); Director, Defense Research and Engineering.
3. International Technology Roadmap for Semiconductors. 2001. Available online at <http://public.itrs.net/> [July 1, 2002].
4. Smith, T. 2001. IBM, AMD unveil terahertz transistor breakthroughs. Available online at <http://www.theregister.co.uk/content/3/23163.html> [July 2, 2002].
5. Intel. 2001. Intel announces breakthrough in chip transistor design. Available online at <http://www.intel.com/pressroom/archive/releases/20011126tech.htm> [July 2, 2002].
6. Tristram, C. 2001. It's time for clockless chips. Available online at <http://www.technologyreview.com/articles/tristram1001.asp> [July 2, 2002].
7. Reed, M.A. 1999. Molecular-scale electronics. Proceedings of the IEEE 87(4): 652–658.
8. Ellenbogen, J.C., and J.C. Love. 2000. Architectures for molecular electronic computers: 1. Logic structures and an adder designed from molecular electronic diodes. Proceedings of the IEEE 88(3): 386–426.
9. Wada, Y. 2001. Prospects for single molecule information processing devices. Proceedings of the IEEE 89(8): 1147–1171.
10. Service, R.F. 2001. Molecules get wired. Science 294(5551): 2442–2443.
11. Meindl, J.D. 1995. Low power microelectronics: Retrospect and prospect. Proceedings of the IEEE 83(4): 619–635.
12. Meindl, J.D. 1996. Physical limits on gigascale integration. Journal of Vacuum Science and Technology B 14(1): 192–195.

13. Meindl, J.D., Q.Chen, and J.A. Davis. 2001. Limits on silicon nanoelectronics for terascale integration. Science 293(5537): 2044–2049.
14. Thompson, S., P. Packan, and M. Bohr. 1998. MOS scaling: Transistor challenges for the 21st century. Intel Technology Journal, 3rd quarter. Available online at <http://developer.intel.com/technology/itj/q31998/articles/art_3.htm> [September 25, 2002].
15. Davis, J.A., V.K. De, and J.D. Meindl. 1998. I. A stochastic wire-length distribution for gigascale integration (GSI)—Part I: Derivation and validation. IEEE Transactions on Electron Devices 45(3): 580–589.
16. Davis, J.A., V.K. De, and J.D. Meindl. 1998. II. A stochastic wire-length distribution for gigascale integration (GSI)—Part II: Application to clock frequency, power dissipation, and chip size estimation. IEEE Transactions on Electron Devices 45(3): 590–597.
17. Davis, J.A., and J.D. Meindl. 1998. Is interconnect the weak link? IEEE Circuits and Devices 14(2): 30–36.
18. Davis, J.A., R. Venkatesan, A. Kaloyeros, M. Beylansky, S.J. Souri, K. Banerjee, K.C. Saraswat, A. Rahman, R. Reif, and J.D. Meindl. 2001. Interconnect limits on gigascale integration (GSI) in the 21st century. Proceedings of the IEEE 89(3): 305–324.
19. Keyes, R.W. 2001. The cloudy crystal ball: Electronic devices for logic. Philosophical Magazine B—Physics of Condensed Matter Statistical Mechanics Electronic Optical and Magnetic Properties 81(9): 1315–1330.
20. Likharev, K.K. 1999. Single-electron devices and their applications. Proceedings of the IEEE 87(4): 606–632.
21. Likharev, K.K. 1999. Single-electron devices and their applications. Proceedings of the IEEE 87(4): 606–632.
22. There is recent evidence that the background charge distributions may be stable for extended lengths of time, Zimmerman, N.M., W.H. Huber, A. Fujiwara, and Y. Takahashi. 2001. Excellent charge offset stability in a Si-based single-electron tunneling transistor. Applied Physics Letters 79(19): 3188–3190.
23. Wolf, S.A., D.D. Awschalom, R.A. Buhrman, J.M. Daughton, S. von Molnár, M.L. Roukes, A.Y. Chtchelkanova, and D.M. Treger, 2001. Spintronics: A spin-based electronics vision for the future. Science 294(5546): 1488–1495.
24. Cowburn, R.P., and M.E. Welland. 2000. Room temperature magnetic quantum cellular automata. Science 287(5457): 1466–1468.
25. Allwood, D.A., G. Xiong, M.D. Cooke, C.C. Faulkner, D. Atkinson, N. Vernier, and R.P. Cowburn. 2002. Submicrometer ferromagnetic NOT gate and shift register. Science 296(5575): 2003–2006.
26. Dimitrakopoulos, C.D., and D.J. Mascaro. 2001. Organic thin-film transistors: A review of recent advances. IBM Journal of Research and Development 45(1): 11–27.
27. Gelinck, G.H., T.C.T. Geuns, and D.M. de Leeuw. 2000. High-performance all-polymer integrated circuits. Applied Physics Letters 77(10): 1487–1489.
28. Allwood, D.A., G. Xiong, M.D. Cooke, C.C. Faulkner, D. Atkinson, N. Vernier, and R.P. Cowburn. 2002. Submicrometer ferromagnetic NOT gate and shift register. Science 296(5575): 2003–2006.
29. Sirringhaus, H., T. Kawase, R.H. Friend, T. Shimoda, M. Inbasekaran, W. Wu, and E.P Woo. 2000. High-resolution inkjet printing of all-polymer transistor circuits. Science 290(5499): 2123–2126.
30. Calvert, P. 2001. Inkjet printing for materials and devices. Chemical Materials 13(10): 3299–3305.
31. Dimitrakopoulos, C.D., and P.R.L. Malenfant. 2002. Organic thin film transistors for large area electronics. Advanced Materials 14(2): 99–117.

32. Tanaka, T., S. Doi, H. Koezuka, A. Tsumura, and H. Fuchigami. 2000. Tanaka, T., S. Doi, H. Koezuka, A. Tsumura, and H. Fuchigami, inventors. Mitsubishi Denki Kabushiki Kaisha and Sumitomo Chemical Company, Limited, assignees. Method of making a field effect transistor. U.S. Patent 6,060,338, May 9.
33. Postma, H.W.Ch., T. Teepen, Z. Yao, M. Grifoni, and C. Dekker. 2001. Carbon nanotube single-electron transistors at room temperature. Science 293(5527): 76–79.
34. Liu, X., C. Lee, C. Zhou, and J. Han. 2001. Carbon nanotube field-effect inverters. Applied Physics Letters 79(20): 3329–3331.
35. Bachtold, A., P. Hadley, T. Nakanishi, and C. Dekker. 2001. Logic circuits with carbon nanotube transistors. Science 294(5545): 1317–1320.
36. Rotman, D. 2002. The nanotube computer. Available online at <http://www.technologyreview.com/articles/rotman0302.asp> [July 2, 2002].
37. Mathews, R.H, J.P. Sage, T.C.L.G. Sollner, S.D. Calawa, C.L. Chen, L.J. Mahoney, P.A. Maki, and K.M. Molvar. 1999. A new RTD-FET logic family. Proceedings of the IEEE 87(4): 596–605.
38. Van Wees, B.J. 1989. Quantum Ballistic and Adiabatic Electron Transport, Studied with Quantum Point Contacts. Thesis, Technische Universiteit Delft, The Netherlands.
39. Altshuler, B.L., P.A. Lee, and R.A. Webb. 1991. Mesoscopic Phenomena in Solids. New York, N.Y.: Elsevier Science.
40. Terabe, K., T. Hasegawa, T. Nakayama, and M. Aono. 2001. Quantum point contact switch realized by solid electrochemical reaction. RIKEN Review 37, Nanotechnology in RIKEN I: 7–8.
41. Granatstein, V.L., R. Parker, and C.M. Armstrong. 1999. Vacuum electronics at the dawn of the twenty-first century. Proceedings of the IEEE 87(5): 702–716.
42. Alles, M., and S. Wilson. 1997. Thin film silicon on insulator: An enabling technology. Semiconductor International 20(4): 67–68.
43. European Space Agency, Space Environment Information System. Available online at <http://www.spenvis.oma.be/spenvis/> [July 2, 2002].
44. Lacoe, R.C., J.V. Osborn, R. Koga, S. Brown, and D.C. Mayer. 2000. Application of hardness-by-design methodology to radiation-tolerant ASIC technologies. IEEE Transactions on Nuclear Science 47(6): 2334–2341.
45. Osborn, J.V., R.C. Lacoe, D.C. Mayer, and G. Yabiku. 1998. Total dose hardness of three commercial CMOS microelectronics foundries. IEEE Transactions on Nuclear Science 45(3): 1458–1463.
46. Lacoe, R.C., J.V. Osborn, R. Koga, S. Brown, and D.C. Mayer. 2000. Application of hardness-by-design methodology to radiation-tolerant ASIC technologies. IEEE Transactions on Nuclear Science 47(6): 2334–2341.
47. Rajchman, J.A. 1961. Computer memories, a survey of the state-of-the-art. Proceedings of the Institute of Radio Engineers 49(1): 104–127.
48. Snider, G.L., A.O. Orlov, I. Amlani, X. Zuo, G.H. Bernstein, C.S. Lent, J.L. Merz, and W. Porod. 1999. Quantum-dot cellular automata: Review and recent experiments. Journal of Applied Physics 85(8): 4283–4285.
49. Snider, G.L., A.O. Orlov, I. Amlani, G.H. Bernstein, C.S. Lent, J.L. Merz, and W. Porod. 1999. Quantum-dot cellular automata. Microelectronic Engineering 47(1–4): 261–263.
50. Cole, T., and J.C. Lusth. 2001. Quantum-dot cellular automata. Progress in Quantum Electronics 25(4): 165–189.
51. Lent, C.S. 2000. Molecular electronics—bypassing the transistor paradigm. Science 288(5471): 1597–1599.
52. Lent, C.S. 2000. Molecular electronics—bypassing the transistor paradigm. Science 288(5471): 1597–1599.
53. Tóth, G., and C.S. Lent. 2001. Quantum computing with quantum-dot cellular automata. Physical Review A 63(5): article number 052315 (9 pages).

54. Two useful algorithms are known, Shor's algorithm for factoring large numbers and Grover's algorithm for searching a database.
55. Felleman, D.J., and D.C. Van Essen. 1991. Distributed hierarchical processing in the primate cerebral cortex. Cerebral Cortex 1(1): 1–47.
56. Abbott, L.F., and S.B. Nelson. 2000. Synaptic plasticity: Taming the beast. Nature Neuroscience Supplement 3(Supp):1178–1183.
57. Kelso, I.A. 1995. Dynamic Patterns: The Self-Organization of Brain and Behavior. Cambridge, Mass.: MIT Press.
58. Freeman, W.J. 2000. Neurodynamics: An Exploration in Mesoscopic Brain Dynamics. New York, N.Y.: Springer.
59. Elman, J.L., E.A. Bates, M.H. Johnson, and A. Karmiloff-Smith. 1996. Rethinking Innateness: A Connectionist Perspective on Development. Cambridge, Mass.: MIT Press.
60. Cauller, L.J. In press. The neurointeractive paradigm: dynamical mechanics and the emergence of higher cortical function. In Theories of Cerebral Cortex, R. Hecht-Neilsen and T. McKenna, eds. San Diego, Calif.: Academic Press.
61. Edelman, G.M., and G. Tononi, G. 2001. A Universe of Consciousness: How Matter Becomes Imagination. New York, N.Y.: Basic Books.
62. Thompson, D.A., and J.S. Best. 2000. The future of magnetic data storage. IBM Journal of Research and Development 44 (3): 311–322.
63. Vettiger, P., M. Despont, U. Drechsler, U. Dürig, W. Häberle, M.I. Lutwyche, H.E. Rothuizen, R. Stutz, R. Widmer, and G.K. Binnig. 2000. The 'Millipede'—more than one thousand tips for future AFM data storage. IBM Journal of Research and Development 44 (3): 323–340.
64. Chen, J., W. Wang, M.A. Reed, A.M. Rawlett, D.W. Price, and J.M. Tour. 2000 Room-temperature negative differential resistance in nanoscale molecular junctions. Applied Physics Letters 77(8): 1224–1226.
65. Kuekes, P.J., R.S. Williams, and J.R. Heath. 2000. Kuekes, P.J., R.S. Williams, and J.R. Heath, inventors. Hewlett-Packard, assignee. Molecular wire crossbar memory. U.S. Patent No. 6.128,214, October 3.
66. Birge, R.R., N.B. Gillespie, E.W. Izaguirre, A. Kusnetzow, A.F. Lawrence, D. Singh, Q.W. Song, E. Schmidt, J.A. Stuart, S. Seetharaman, and K.J. Wise. 1999. Biomolecular electronics: Protein-based associative processors and volumetric memories. Journal of Physical Chemistry 103B(49): 10746–10766.
67. Ledentsov, N.N., M. Grundmann, F. Heinrichsdorff, D. Bimberg, V.M. Ustinov, A.E. Zhukov, M.V. Maximov, Z.I. Alferov, and J.A. Lott. 2000. Quantum dot heterostructure lasers. IEEE Journal of Selected Topics in Quantum Electronics 6(3): 439–451.
68. See, for example, N. Holonyak. 1997. The semiconductor laser: A thirty-five year perspective. Proceedings of the IEEE 85(11): 1678–1693.
69. Felix, C.L., W.W. Bewley, I. Vurgarfman, R.E. Bartolo, D.W. Stokes, J.R. Meyer, M.J. Yang, H. Lee, R.J. Menna, R.U. Martinelli, D.Z. Garbuzov, J.C. Connolly, M. Maiorov, A.R. Sugg, and G.H. Olsen. 2001. Mid-infrared W quantum-well lasers for noncryogenic continuous-wave operation. Applied Optics 40(6): 806–811.
70. Capasso, F., C. Gmachl, R. Paiella, A. Tredicucci, A.L. Hutchinson, D.L. Sivco, J.N. Baillargeon, A.Y. Cho, and H.C. Liu. 2000. New frontiers in quantum cascade lasers and applications. IEEE Journal of Selected Topics in Quantum Electronics 6(6): 931–947.
71. Gao, H., and W.D. Nix. 1999. Surface roughening of heteroepitaxial thin films. Annual Review of Materials Science 29: 173–209; A. Shchukin and D. Bimberg. 1999. Spontaneous ordering of nanostructures on crystal surfaces. Reviews in Modern Physics 71(4): 1125–1171.
72. Wang, R.H., A. Stintz, P.M. Varangis, T.C. Newell, H. Li, K.J. Malloy, and L.F. Lester. 2001. Room-temperature operation of InAs quantum-dash lasers on InP (001). IEEE Photonics Technology Letters 13(8): 767–769.

73. Coldren, L.A. 2000. Monolithic tunable diode lasers. IEEE Journal of Selected Topics in Quantum Electronics 6(6): 988–999.
74. See, for example, Erdogan, T., E.J. Friebele, and R. Kashyap. 2000. Bragg Gratings, Photosensitivity, and Poling in Glass Waveguides. Presented at the Topical Meeting on Bragg Grating, Photosensitivity, and Poling in Glass Waveguides, September 23–25, 1999, at Stuart, Fla. Washington, D.C.: Optical Society of America.
75. Towe, E., R.F. Leheney, and A. Yang. 2000. A historical perspective of the development of the vertical-cavity surface-emitting laser. IEEE Journal of Selected Topics in Quantum Electronics 6(6): 1458–1464.
76. Chou, M.H., I. Brener, M.M. Fejer, E.E. Chaban, and S.B. Christman. 1999. 1.5 mm m-band wavelength conversion based on cascaded second-order nonlinearity in LiNbO3 waveguides. IEEE Photonics Technology Letters 11(6): 653–655.
77. Painter, O., K. Srinivasan, J.D. O'Brien, A. Scherer, and P.D. Dapkus. 2001. Tailoring of the resonant mode properties of optical nanocavities in two-dimensional photonic crystal slab waveguides. Journal of Optics A: Pure and Applied Optics 3(6): S161–S170.
78. Yablonovitch, E. 2001. Photonic crystals: Semiconductors of light. Scientific American 285(6): 46–55.
79. Pendry, J.B., A.J. Holden, W.J. Stewart, and I. Youngs. 1996. Extremely low frequency plasmons in metallic mesostructures. Physical Review Letters 76(25): 4773–4776.
80. Shelby, R.A., D.A. Smith, and S. Schultz. 2001. Experimental verification of a negative index of refraction. Science 292(5514): 77–79.
81. Chang-Hasnain, C., E. Vail, and M. Wu. 1996. Widely-tunable micro-mechanical vertical cavity lasers and detectors. Pp. C43–44 in Digest of the IEEE/LEOS 1996 Summer Topical Meetings – Advanced Applications of Lasers in Materials and Processing. New York, N.Y.: Institute of Electrical and Electronics Engineers, Inc.
82. Bishop, D., P. Gammel, and C.R. Gile. 2001. The little machines that are making it big. Physics Today 54(10): 38–44.
83. Hornbeck, L. 1995. Digital Light Processing and MEMS: Timely Convergence for a Bright Future. Available online at <http://www.dlp.com/dlp_technology/images/dynamic/white_papers/107_DLP_MEMS_Overview.pdf> [July 2, 2002].
84. Johnson, C. 2002. The wireless monster. The Industry Standard, January 10. Available online at <http://www.thestandard.com/article/0,1902,8551,00.html> [July 2, 2002].
85. Yao, J.J., S.T. Park, and J. DeNatale. 1998. High tuning-ratio MEMS-based tunable capacitors for RF communications applications. Technical Digest: Solid-State Sensor and Actuator Workshop. Cleveland, Ohio: Transducer Research Foundation.
86. Yao, Z.J., S. Chen, S. Eshelman, D. Denniston, and C. Goldsmith. 1999. Micromachined low-loss microwave switches. Journal of Microelectromechanical Systems 8(2): 129–34.
87. Goldsmith, C. 2001. RF MEMS. Paper presented to the 7th World Micromachine Summit, Frieburg, Germany, April 30–May2.
88. Vaughan, C.R. 2001. Defining specifications for consumer focused MEMS and the cost performance trade-offs. Paper presented at the 2nd International Conference COTS MEMS 2001—Advances in Application of Integrated Commercial Off-The-Shelf MicroElectroMechanical Systems. Boston, Mass., November 30.
89. John Wiley & Sons. 2000. Technical Insights, R-263: Optical MEMS: Worldwide Markets For a Strategic and Convergent Technology. New York, N.Y.: John Wiley & Sons, Inc.
90. Wicht Technologie Consulting. 2002. The Market for RF MEMS 2002–2007: Market Analysis and Technology Roadmap for RF MEMS Components and Applications (Draft). Available online at <http://www.wtc-consult.de/download/rfmems/rfabstra.PDF> [July 2, 2002].
91. National Research Council. 2001. Embedded, Everywhere: A Research Agenda for Networked Systems of Embedded Computers. Washington, D.C.: National Academy Press.

92. Agee, F. 2001. AFRL Overview to the NRC on Nano Technologies. Briefing by Forrest (Jack) Agee, Air Force Office of Scientific Research, to the Committee on Implications of Emerging Micro and Nano Technologies, Holiday Inn, Fairborn, Ohio, October 2.
93. Kahn, J.M., R.H. Katz, and K.S.J. Pister. 1999. Next century challenges: Mobile networking for smart dust. Available online at <http://robotics.eecs.berkeley.edu/~pister/publications/1999/mobicom_99.pdf> [August 22,2002].
94. National Research Council. 1997. Microelectromechanical Systems: Advanced Materials and Fabrication Methods. Washington, D.C.: National Academy Press.
95. Agee, F. 2001. AFRL Overview to the NRC on Nano Technologies. Briefing by Forrest (Jack) Agee, Air Force Office of Scientific Research, to the Committee on Implications of Emerging Micro and Nano Technologies, Holiday Inn, Fairborn, Ohio, October 2.
96. Huang, M.H., S. Mao, H. Feick, H.Q. Yan, Y.Y. Wu, H. Kind, E. Weber, R. Russo, and P.D. Yang. 2001. Room-temperature ultraviolet nanowire nanolasers. Science 292(5523): 1897–1899. Crystalline zinc oxide (ZnO) nanowires were demonstrated in the ultraviolet at 385 nm with a linewidth of less than 0.3 nm.
97. Agee, F. 2001. AFRL Overview to the NRC on Nano Technologies. Briefing by Forrest (Jack) Agee, Air Force Office of Scientific Research, to the Committee on Implications of Emerging Micro and Nano Technologies, Holiday Inn, Fairborn, Ohio, October 2.
98. Guenther, R.D. 2001. NRC Committee on Implications of Emerging Micro and Nano Technologies. Briefing by Robert D. Guenther, Army Research Office-IPA/Duke University, to the Committee on Implications of Emerging Micro and Nano Technologies, National Academy of Sciences, Washington, D.C., August 16, 2001.
99. Specifically, wavefront measurement advances have resulted in improved characterization of human vision, including the aberrations of the entire human optical path. This development of integrated adaptive optics (AO) and sodium-layer laser guide star (LGS) systems for use on large astronomical telescopes has been a boon to this effort. See Max, C.E., S.S. Olivier, H.W. Friedman, K. An, K. Avicola, B.V. Beeman, H.D. Bissinger, J.M. Brase, G.V. Erbert, D.T. Gavel, K. Kanz, M.C. Liu, B. Macintosh, K.P. Neeb, J. Patience, and K.E. Waltjen. 1997. Image improvement from a sodium-layer laser guide star adaptive optics system. Science 277(5332): 1649–1652; and Avicola, K., J.T. Salmon, J. Brase, K. Waltjen, R. Presta, and K.S. Bach. 1992. High frame-rate large field wavefront sensor. Proceedings of the Laser Guide Star Adaptive Optics Workshop. Livermore, Calif.: Lawrence Livermore National Laboratory.
100. Yoon, G.Y., and D.R. Williams. 2002. Visual performance after correcting the monochromatic and chromatic aberrations of the eye. Journal of the Optical Society of America A: Optics Image Science and Vision 19(2): 266–275.
101. Agee, F. 2001. AFRL Overview to the NRC on Micro and Nano Technologies. Briefing by Forrest (Jack) Agee, Air Force Office of Scientific Research, to the Committee on Implications of Emerging Micro and Nano Technologies, National Academy of Sciences, Washington, D.C., August 16.
102. Daniel, D. 2001. Air Force Science and Technology Overview. Briefing by Don Daniel, Deputy Assistant Secretary of the Air Force (Science, Technology, and Engineering), to the Committee on Implications of Emerging Micro and Nano Technologies, National Academy of Sciences, Washington, D.C., August 15.
103. Brown, G.J. 2001. Nanoelectronics and Nanomaterials for Sensors. Briefing by Gail J. Brown, Air Force Research Laboratory, Materials and Manufacturing Directorate, to the Committee on Implications of Emerging Micro and Nano Technologies, Holiday Inn, Fairborn, Ohio, October 2.
104. Sobolewski, R., D. P. Butler, and Z. Celik-Butler. 2001. Cooled and uncooled infrared detectors based on yttrium barium copper oxide. Pp. 204–214 in Smart Optical Inorganic Structures and Devices, Proceedings of SPIE Volume 4318. S.P. Asmontas and J. Gradauskas, eds. Bellingham, Wash.: The International Society for Optical Engineering.

105. Maranowski, K.D., J.M. Peterson, S.M. Johnson, J.B. Varesi, A.C. Childs, R.E. Bornfreund, A.A. Buell, W.A. Radford, T.J. de Lyon, and J.E. Jensen. 2001. MBE growth of HgCdTe on silicon substrates for large format MWIR focal plane arrays. Journal of Electronic Materials 30(6): 619–622.
106. AFOSR has been funding this work. Brown, G.J. 2001. Nanoelectronics and Nanomaterials for Sensors. Briefing by Gail J. Brown, Air Force Research Laboratory, Materials and Manufacturing Directorate, to the Committee on Implications of Emerging Micro and Nano Technologies, Holiday Inn, Fairborn, Ohio, October 2. A superlattice in this context consists of alternating layers of different semiconductor materials, each several nanometers thick.
107. Mohseni, H., M. Razeghi, G.J. Brown, and Y.S. Park. 2001. High-performance InAs/GaSb superlattice photodiodes for the very long wavelength infrared range. Applied Physics Letters 78(15): 2107–2109.
108. Klappenberger, F., A.A. Ignatov, S. Winnerl, E. Schomburg, W. Wegscheider, K.F. Renk, and M. Bichler. 2001. Broadband semiconductor superlattice detector for THz radiation. Applied Physics Letters 78(12): 1673–1675.
109. Bahl, I.J., and P. Bhartia. 1988. Microwave Solid State Circuit Design. New York, N.Y.: John Wiley & Sons.
110. Schulman, J.N., K.S. Holabird, D.H. Chow, H.L. Dunlap, S. Thomas, and E.T. Croke. 2002. Temperature dependence of Sb-heterostructure millimetre-wave diodes. Electronics Letters 38(2): 94–95.
111. Schulman, J.N., and D.H. Chow. 2000. Sb-heterostructure interband backward diodes. IEEE Electron Device Letters 21(7): 353–355.
112. Leheny, R., 2001. Implications of Emerging Micro and Nano Technologies. Briefing by Robert Leheny, Director, DARPA, to the Committee on Implications of Emerging Micro and Nano Technologies, National Academy of Sciences, Washington, D.C., on August 16, 2001.
113. Centre Suisse d'Electronique et de Microtechnique SA. 1994. Data sheets ASEM02-S and ASEM02-T/6. Neuchatel, Switzerland: Centre Suisse d'Electronique et de Microtechnique SA.
114. Kukkonen, C.A. 1995. NASA Vision and Implementation Approach for Advanced Technologies in Space. Presentation by Carl A. Kukkonen, Jet Propulsion Laboratory, Presentation to the 2nd Round Table on Micro and Nano Technologies for Space, European Space Agency ESTEC, Nordwijk, The Netherlands, October 15–17, 2002.
115. For a broad review of many automotive sensors, see Fleming, W.J. 2001. Overview of automotive sensors. IEEE Sensors Journal 1(4): 296–308.
116. Kubena, R.L., D.J. Vickers-Kirby, R.J. Joyce, and F.P. Stratton. 1999. A new tunneling-based sensor for inertial rotation rate measurement. Journal of Micromechanical Systems 8(4): 439–447.
117. Honeywell. 1995. HMC2003 Three-Axis Magnetic Sensor Hybrid Data Sheet, Rev. C 10/99. Available online at <http://www.ssec.honeywell.com/magnetic/datasheets/hmc2003.pdf> [July 2, 2002].
118. Nonvolatile Electronics. 1995. Rapid Prototype Integrated GMR Magnetic Sensors Data Sheet, GMR Magnetic Bridge Sensor—NVSB Data Sheet, and Integrated GMR Magnetic Sensors—NVSI Data Sheet. Eden Prairie, Minn.: Nonvolatile Electronics, Inc.
119. Wickenden, D.K., T.J. Kistenmacher, R. Osiander, S.A. Ecelberger, R.B. Givens, and J.C. Murphy. 1997. Development of Miniature Micromagnets. Johns Hopkins APL Technology Digest 18(2): 271–278.
120. National Research Council. 2001. Opportunities in Biotechnology for Future Army Applications. Washington, D.C.: National Academy Press.
121. Thayer, A.M. 2001. Nanotech offers some there, there. Chemical & Engineering News 79(48): 13–16.

122. Ilic, B., D. Czaplewski, M. Zalalutdinov, H.G. Craighead, P. Neuzil, C. Campagnolo, and C. Batt. 2001. Single cell detection with micromechanical oscillators. Journal of Vacuum Science and Technology B 19(6): 2825–2828.
123. Cooper, M.A., F.N. Dultsev, T. Minson, V.P. Ostanin, C. Abell, and D. Klenerman. 2001. Direct and sensitive detection of a human virus by rupture event scanning. Nature Biotechnology 19(9): 833–837.
124. Kong, J., N.R. Franklin, C.W. Zhou, M.G. Chapline, S. Peng, K.J. Cho, and H.J. Dai. 2000. Nanotube molecular wires as chemical sensors. Science 287(5453): 622–625.
125. Horworka, S., S. Cheley, and H. Bayley. 2001. Sequence-specific detection of individual DNA strands using engineered nanopores. Nature Biotechnology 19(7): 636–639.
126. Park, S.J., T.A. Taton, and C.A. Mirkin. 2002. Array-based electrical detection of DNA with nanoparticle probes. Science 295(5559): 1503–1506.
127. Edelstein, R.L., C.R. Tamanaha, P.E. Sheehan, M.M. Miller, D.R. Baselt, L.J. Whitman, and R.J. Colton. 2000. The BARC biosensor applied to the detection of biological warfare agents. Biosensors and Bioelectronics 14(10–11): 805–813.
128. Guenther, R.D. 2001. NRC Committee on Implications of Emerging Micro and Nano Technologies. Briefing by Robert D. Guenther, Army Research Office-IPA/Duke University, to the Committee on Implications of Emerging Micro and Nano Technologies, National Academy of Sciences, Washington, D.C., August 16.
129. Udd, E., W.L. Schulz, J.M. Seim, A. Trego, E. Haugse, and P.E. Johnson. 2000. Transversely loaded fiber optic grating strain sensors for aerospace applications. Pp. 96–104 in Nondestructive Evaluation of Aging Aircraft, Airports, and Aerospace Hardware IV, Proceedings of SPIE Volume 3994. A.K. Mal, ed. Bellingham, Wash.: The International Society for Optical Engineering.
130. Belk, J.H., and E.V. White. 1999. Belk, J.H., and E.V. White, inventors. McDonnell Douglas Corporation, assignee. Remotely Interrogatable Apparatus and Method for Detecting Defects in Structural Members. U.S. Patent 5,969,260, October 19.
131. Krantz, D.G., J. Belk, P.J. Biermann, and P. Troyk. 2000. Project summary: applied research on remotely-queried embedded microsensors. Pp. 110–121 in Smart Structures and Materials 2000: Smart Electronics and MEMS, Proceedings of SPIE Volume 3990. V.K. Varadan, ed. Bellingham, Wash.: The International Society for Optical Engineering.
132. Kim, N.P., M.J. Holland, M.H. Tanielian, and R. Poff. 2000. MEMS sensor multi-chip module assembly with TAB carrier—Pressure belt for aircraft flight testing. Pp. 689–696 in Proceedings of the Electronic Components and Technology Conference 2000. New York, N.Y.: IEEE.
133. Agee, F. 2001. AFRL Overview to the NRC on Nano Technologies. Briefing by Forrest (Jack) Agee, Air Force Office of Scientific Research, to the Committee on Implications of Emerging Micro and Nano Technologies, Holiday Inn, Fairborn, Ohio, October 2.
134. Defense Advanced Research Projects Agency. 2002. Chip Scale Atomic Clock Overview. Available online at <http://www.darpa.mil/mto/csac/overview/index.html> [July 2, 2002].
135. Holzwarth, R., T. Udem, T.W. Hänsch, J.C. Knight, W.J. Wadsworth, and P.S.J. Russell. 2000. Optical frequency synthesizer for precision spectroscopy. Physical Review Letters 85(11): 2264–2267.
136. Kahn, J.M., R.H. Katz, and K.S.J. Pister. 1999. Next century challenges: Mobile networking for Smart Dust. Available online at <http://robotics.eecs.berkeley.edu/~pister/publications/1999/mobicom_99.pdf> [August 22,2002].
137. Defense Advanced Research Projects Agency. 2002. Smart Dust Project Summary. Available online at <http://www.darpa.mil/mto/mems/summaries/projects/individual_57.html> [July 2, 2002].
138. Nyquist, R.M., A.S. Eberhardt, L.A. Silks, Z. Li, X. Yang, and B.I. Swanson. 2000. Characterization of self-assembled monolayers for biosensor applications. Langmuir 16(4): 1793–1800.

139. Hou, Z.Z, S. Dante, N.L. Abbott, and P. Stroeve. 1999. Self-assembled monolayers on (111) textured electroless gold. Langmuir 15(8): 3011–3014.
140. Bachand, G.D., R.K. Soong, H.P. Neves, A. Olkhovets, H.G. Craighead, and C.D. Montemagno. 2001. Precision attachment of individual F-1-ATPase biomolecular motors on nanofabricated substrates. Nano Letters 1(1): 42–44.
141. Mann, S., W. Shenton, M. Li, S. Connolly, and D. Fitzmaurice. 2000. Biologically programmed nanoparticle assembly. Advanced Materials 12(2): 147–150.
142. Lee, K.B., S.J. Park, C.A. Mirkin, J.C. Smith, and M. Mrksich. 2002. Protein nanoarrays generated by dip-pen nanolithography. Science 295(5560): 1702–1705.
143. Guzman-Jimenez, I.Y., K.H. Whitmire, K. Umezama-Vizzini, R. Colorado, J.W. Do, A. Jacobson, T.R. Lee, S.H. Hong, and C.A. Mirkin. 2001. Self-assembly of organometallic clusters onto the surface of gold. Thin Solid Films 401(1–2): 131–137.
144. Weinberger, D.A., S.G. Hong, C.A. Mirkin, B.W. Wessels, and T.B. Higgins. 2000. Combinatorial generation and analysis of nanometer- and micrometer-scale silicon features via "dip-pen" nanolithography and wet chemical etching. Advanced Materials 12(21): 1600–1603.
145. Doshi, D.A., N.K. Huesing, M.C. Lu, H.Y. Fan,Y.F. Lu, K. Simmons-Potter, B.G. Potter, A.J. Hurd, and C.J. Brinker. 2000. Optically, defined multifunctional patterning of photosensitive thin-film silica mesophases. Science 290(5489): 107–111.
146. Lu, Y., Y. Yang, A. Sellinger, M. Lu, J. Huang, H. Fan, R. Haddad, G. Lopez, A.R. Burns, D.Y. Sasaki, J. Shelnutt, and C.J. Brinker. 2001. Self-assembly of mesoscopically ordered chromatic polydiacetylene/silica nanocomposites. Nature 410(6831): 913–917.
147. Oakley, C., N.A.F. Jaeger, and D.M. Brunette. 1997. Sensitivity of fibroblasts and their cytoskeletons to substratum topographies: Topographic guidance and topographic compensation by micromachined grooves of different dimensions. Experimental Cell Research 234(2): 413–424.
148. Fisher, A.B., S. Chien, A.I. Barakat, and R.M. Nerem. 2001. Endothelial cellular response to altered shear stress. American Journal of Physiology—Lung Cellular and Molecular Physiology 281(3): L529–L533.
149. Girard, P.R., and R.M. Nerem. 1995. Shear stress modulates endothelial-cell morphology and F-actin organization through the regulation of focal adhesion-associated proteins. Journal of Cellular Physiology 163(1): 179–193.
150. Gray, B.L., D.K. Lieu, S.D. Collins, R.L. Smith, and A.I. Barakat. 2002. Microchannel platform for the study of endothelial cell shape and function. Biomedical Microdevices 4(1): 9–16.
151. Pullar, C.E., R.R. Isseroff, and R. Nuccitelli. 2001. Cyclic AMP-dependent protein kinase A plays a role in the directed migration of human keratinocytes in a DC electric field. Cell Motility and the Cytoskeleton 50(4): 207–217.
152. Farboud, B., R. Nuccitelli, I.R. Schwab, and R.R. Isseroff. 2000. DC electric fields induce rapid directional migration in cultured human corneal epithelial cells. Experimental Eye Research 70(5): 667–673.
153. Marder, E., and R. L. Calabrese. 1996. Principles of rhythmic motor pattern generation. Physiological Reviews 76(3): 687–717.
154. Kiehn, O., O. Kjaerulff, M.C. Tresch, and R.M. Harris-Warrick. 2000. Contributions of intrinsic motor neuron properties to the production of rhythmic motor output in the mammalian spinal cord. Brain Research Bulletin 53(5): 649–659.
155. Woodhouse, G., L. King, L. Wieczorek, P. Osman, and B. Cornell. 1999. The ion channel switch biosensor. Journal of Molecular Recognition 12(5): 328–334.
156. Pace, R.J., V.L. Braach-Maksvytis, L.G. King, P.D. Osman, B. Raguse, L. Wieczorek, and B.A. Cornell. 1998. The gated ion channel biosensor—a functioning nanomachine. Pp. 50–59 in Methods for Ultrasensitive Detection, Proceedings of SPIE Volume 3270. B.L. Fearey, ed. Bellingham, Wash.: The International Society for Optical Engineering.
157. Adleman, L.M. 1998. Computing with DNA. Scientific American 279(2): 54–61.

158. Howe, R.T., and R.S. Muller. 1986. Resonant-microbridge vapor sensor. IEEE Transactions on Electron Devices 33(4): 499–506.
159. Fritz, J., M.K. Baller, H.P. Lang, H. Rothuizen, P. Vettiger, E. Meyer, H.J. Guntherodt, C. Gerber, and J.K. Gimzewski. 2000. Translating biomolecular recognition into nanomechanics. Science 288(5464): 316–318.
160. Baller, M.K., H.P. Lang, J. Fritz, C. Gerber, J.K. Gimzewski, U. Drechsler, H. Rothuizen, M. Despont, P. Vettiger, F.M. Battiston, J.P. Ramseyer, P. Fornaro, E. Meyer, H.J. Guntherodt. 2000. A cantilever array-based artificial nose. Ultramicroscopy 82(1–4): 1–9.
161. Yoo, K., J.-L.A. Yeh, N.C. Tien, C. Gibbons, Q. Su, W. Cui, and R.N. Miles. 2001. Fabrication of a biomimetic corrugated polysilicon diaphragm with attached single crystal silicon proof masses. Available online at <http://www.me.binghamton.edu/miles/current%20research/Trans01final.pdf> [July 3, 2002].
162. Byl, C., D.W. Howard, S.D. Collins, and R.L. Smith. 1999. Micromachined, Multi-Axis Accelerometer with Liquid Proof Mass, Final Report 1998–99 for MICRO Project 98-145. Available online at <http://www.ucop.edu/research/micro/98_99/98reports.html> [July 2, 2002].
163. Petch, N.J. 1953. The cleavage strength of polycrystals. Journal of the Iron and Steel Institute 173: 25–28.
164. Lowe, T., 2001. Nanometals and Air Force Technology Development. Briefing by Terry C. Lowe, CEO, Metallicum, LLC, to the Committee on Implications of Emerging Micro and Nano Technologies, National Academy of Sciences, Irvine, Calif., December 19, 2001.
165. National Science and Technology Council. 1999. Nanotechnology Research Directions: IWGN Workshop Report, September. Available online at <http://itri.loyola.edu/nano/IWGN.Research.Directions/> [July 3, 2002].
166. National Research Council. 2001. A Summary of the Workshop on Structural Nanomaterials, June 20–21. Washington, D.C.: National Academy Press.
167. Lowe, T. 2001. Nanometals and Air Force Technology Development. Briefing by Terry C. Lowe, CEO, Metallicum, LLC, to the Committee on Implications of Emerging Micro and Nano Technologies, National Academy of Sciences, Irvine, Calif., December 19, 2001.
168. Yu, M.F, B.S. Files, S. Arepalli, and R.S. Ruoff. 2000. Tensile loading of ropes of single wall carbon nanotubes and their mechanical properties. Physical Review Letters 84(24): 5552–5555.
169. Walters, D.A., L.M. Ericson, M.J. Casavant, J. Liu, D.T. Colbert, K.A. Smith, and R.E. Smalley. 1999. Elastic strain of freely suspended single-wall carbon nanotube ropes. Applied Physics Letters 74(25): 3803–3805.
170. Wang, Z.L., R.P. Gao, Z.W. Pan, and Z.R. Dai. 2001. Nano-scale mechanics of nanotubes, nanowires, and nanobelts. Advanced Engineering Materials 3(9): 657–661.
171. See, for example, Girshick, S.L., W.W. Gerberich, J.V.R. Heberlein, and P.H. McMurry. 2001. Nanotechnology Highlight: Microfabrication with focused beams of nanoparticles. Available online at <http://www-unix.oit.umass.edu/~nano/NewFiles/FN15_Minn_NH.pdf> [July 3, 2002].
172. National Science and Technology Council. 1999. Nanotechnology Research Directions: IWGN Workshop Report, September. Available online at <http://itri.loyola.edu/nano/IWGN.Research.Directions/> [July 3, 2002].
173. Misra, A. and H. Kung. 2001. Deformation behavior of nanostructured metallic multilayers. Advanced Engineering Materials 3(4): 217–222.
174. Berber, S., Y.K Kwon, and D. Tománek. 2000. Unusually high thermal conductivity of carbon nanotubes. Physical Review Letters 84(20): 4613–4616.
175. Naval Research Laboratory Multifunctional Materials Branch; see <http://mstd.nrl.navy.mil/6350/6350.html> [July 3, 2002].
176. Neves, B.R.A., M.E. Salmon, E.B. Troughton, and P.E. Russell. 2001. Self-healing on OPA self-assembled monolayers. Nanotechnology 12(3): 285–289.

177. See, for example, Guckel, H. 2000. Built-in strain in polysilicon: Measurement and application to sensor fabrication. Pp. 3–12 in Materials Science of Microelectromechanical Systems (MEMS) Devices II, MRS Symposium Proceedings Volume 605. M.P. DeBoer, A.H. Heuer, S.J. Jacobs, and E. Peeters, eds. Warrendale, Pa.: Materials Research Society.
178. Habermehl, S., A.K. Glenzinski, W.M. Halliburton, and J.J. Sniegowski. 2000. Properties of low residual stress silicon oxynitrides used as a sacrificial layer. Pp. 49–54 in Materials Science of Microelectromechanical Systems (MEMS) Devices II, MRS Symposium Proceedings Volume 605. M.P. DeBoer, A.H. Heuer, S.J. Jacobs, and E. Peeters, eds. Warrendale, Pa.: Materials Research Society.
179. See, for example, Chasiotis, I., and W.G. Knauss. 2001. The influence of fabrication governed by surface conditions on the mechanical strength of thin film materials. Pp. EE2.2.1–EE2.2.6 in Materials Science of Microelectromechanical Systems (MEMS) Devices III, MRS Symposium Proceedings Volume 657. M. DeBoer, M. Judy, H. Kahn, and S.M. Spearing, eds. Warrendale, Pa.: Materials Research Society.
180. See, for example, Adams, P.M., R.E. Robertson, R.C. Cole, D. Hinkley, and G. Radhakrishnan. 2000. Investigation of the deposition and integration of hard coatings for moving MEMS applications. Pp. 123–128 in Materials Science of Microelectromechanical Systems (MEMS) Devices II, MRS Symposium Proceedings Volume 605. M.P. DeBoer, A.H. Heuer, S.J. Jacobs, and E. Peeters, eds. Warrendale, Pa.: Materials Research Society.
181. Mastrangelo, C.H. 2000. Suppression of stiction in MEMS. Pp. 105–116 in Materials Science of Microelectromechanical Systems (MEMS) Devices II, MRS Symposium Proceedings Volume 605. M.P. de Boer, A.H. Heuer, S.J. Jacobs, and E. Peeters, eds. Warrendale, Pa.: Materials Research Society.
182. See, for example, Louchet, F., J.J. Blandin, and M. Véron. 2001. In situ transmission electron microscopy study of the strength and stability of nanoscaled structural materials. Advanced Engineering Materials 3(8): 608–612.
183. Huang, A., C. Folk, C.M. Ho, Z. Liu, W.W. Chu, Y. Xu, and Y.C. Tai. 2001. Gryphon M3 system: Integration of MEMS for flight control. Pp. 85–94 in MEMS Components and Applications for Industry, Automobiles, Aerospace, and Communication, Proceedings of the SPIE Volume 4559. H. Helvajian, S.W. Janson, and F. Larmer, eds. Bellingham, Wash.: The International Society for Optical Engineering.
184. Huang, A., C. Folk, C.M. Ho, Z. Liu, W.W. Chu, Y. Xu, and Y.C. Tai. 2001. Gryphon M3 system: Integration of MEMS for flight control. Pp. 85–94 in MEMS Components and Applications for Industry, Automobiles, Aerospace, and Communication, Proceedings of the SPIE Volume 4559. H. Helvajian, S.W. Janson, and F. Larmer, eds. Bellingham, Wash.: The International Society for Optical Engineering.
185. Grasmeyer, J.M. and M.T. Keennon. 2001. Development of the Black Widow Micro Air Vehicle, AIAA Technical Paper 2001-0127. Reston, Va.: American Institute of Aeronautics and Astronautics.
186. Lewis, D.H., S.W. Janson, R.B. Cohen, and E.K. Antonsson. 1999. Digital micropropulsion. Pp. 517–522 in Technical Digest of MEMS '99: Twelfth IEEE International Conference on Micro Electro Mechanical Systems. New York, N.Y.: IEEE.
187. Youngner, D.W., S.T. Lu, E. Choueiri, J.B. Neidert, R.E. Black, K.J. Graham, D. Fahey, R. Lucus, and X. Zhu. 2000. MEMS Mega-pixel Micro-thruster Arrays for Small Satellite Stationkeeping. Available online at <http://www.sdl.usu.edu/conferences/smallsat/proceedings/14/tsx/x-2.pdf> [July 3, 2002].
188. Rossi, C., D. Esteve, N. Fabre, T. Do Conto, V. Conedera, D. Dilhan, and Y. Guelou. 1999. A new generation of MEMS based microthrusters for microspacecraft applications. Pp. 201–209 in Proceedings of the 2nd International Conference on Integrated Micro/Nanotechnology for Space Applications, Volume 1. Los Angeles, Calif.: The Aerospace Corporation.

189. Martin, D.H. 2000. Communication Satellites, 4th Edition. Reston, Va.: American Institute of Aeronautics and Astronautics.
190. Iannazzo, S. 1993. A survey of the present status of vacuum microelectronics. Solid State Electronics 36(3): 301–320.
191. Jensen, K.L., R.H. Abrams, and R.K. Parker. 1998. Field emitter array development for high frequency applications. Journal of Vacuum Science Technology B 16(2): 749–753.
192. Spindt, C.A., C.E. Holland, P.R. Schwoebel, and I. Brodie. 1998. Field emitter array development for microwave applications: II. Journal of Vacuum Science & Technology B 16(2): 758–761.
193. Jo, S.H., K.W. Jung, Y.J. Kim, S.H. Ahn, J.H. Kang, H.S. Han, B.G. Lee, J.C. Cha, S.J. Lee, S.Y. Park, C.G. Lee, J.H. You, N.S. Lee, and J.M. Kim. 2001. Carbon nanotube cathode with low operating voltage. Pp. 31–32 in IVMC 2000: Proceedings of the 14th International Vacuum Microelectronics Conference. New York, N.Y.: IEEE.
194. Spindt, C.A. 1992. Microfabricated field-emission and field-ionization sources. Surface Science 266(1–3): 145–154.
195. Kwon, S.J., Y.H. Shin, D.M. Aslam, and J.D. Lee. 1998. Field emission properties of the polycrystalline diamond film prepared by microwave-assisted plasma chemical vapor deposition. Journal of Vacuum Science & Technology B 16(2): 712–715.
196. Fursey, G.N., L.A. Shirochin, and L.M. Baskin. 1997. Field-emission processes from a liquid-metal surface. Journal of Vacuum Science & Technology B 15(2): 410–421.
197. Taylor, G. 1964. Disintegration of water drops in electric field. Proceedings of the Royal Society of London Series A—Mathematical and Physical Sciences 280(138): 383–397.
198. Mitterauer, J. 1991. Miniaturized liquid-metal ion sources (MILMIS). IEEE Transactions on Plasma Science 19(5): 790–799.
199. Mitterauer, J. 1992. Prospects of liquid metal ion thrusters for electric propulsion. Paper 105 in Proceedings of the AIDAA/AIAA/DGLR/JSASS 22nd International Electric Propulsion Conference, Volume 2. Pisa, Italy: Centrospazio.
200. Spindt, C.A. 1992. Microfabricated field-emission and field-ionization sources. Surface Science 266(1–3): 145–154.
201. Hoyt, R.P. 2000. Design and Simulation of a Tether Boost Facility for LEO to GTO Transport, AIAA paper 2000-3866. Reston, Va.: American Institute of Aeronautics and Astronautics.
202. Hoyt, R.P., and C. Uphoff. 1999. Cislunar Tether Transport System, AIAA paper 99-2690. Reston, Va.: American Institute of Aeronautics and Astronautics.

4

Enabling Manufacturing Technologies

Much of what is called manufacturing science applies to a very wide range of products and size scales. Manufacturability and quality control, which are closely related, vary widely with the type and specifics of a product, whether for the military or for mass markets. Production of microscale surface finishes on a material (for example, a polished metal surface on an aircraft engine turbine blade) is certainly different from making parts for MEMS devices on a similar size scale (for example, a microaccelerometer with submicrometer feature sizes). Similarly, the production of nanocrystalline materials is generally very different from the manufacture of possible molecular electronic devices. In this chapter, the committee considers the general aspects, along with some particular aspects, of manufacturing products whose performance depends on structure, material, or chemistry on the micro- and nanoscales.

The materials, components, subsystems, systems, and platforms used by the military are mainly purchased from industry. Rarely is military hardware manufactured by the Department of Defense. Hence, industry-based manufacture of what the military uses is of central importance. This will be no less true for hardware made by micro- and nanotechnologies than it is for the alloys, antennas, radars, missiles, and airplanes that are now being employed by the military. The role of various communities in acquiring, maintaining, and employing military hardware is summarized in Figure 4-1.

FABRICATION (PATTERNING) APPROACHES

Integrated circuit manufacturing is a top-down process where the starting point is a flat wafer onto which patterns are defined and created by both additive

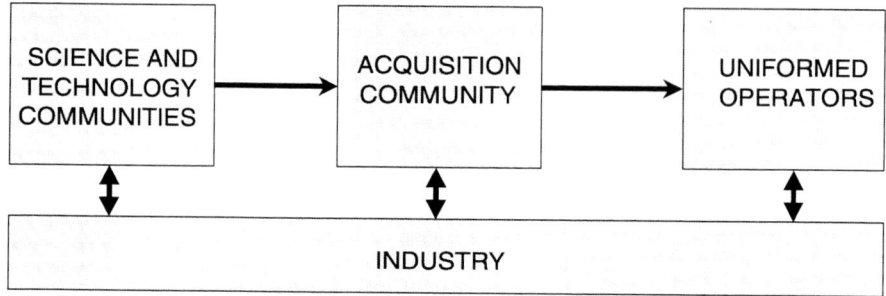

FIGURE 4-1 Communities needed for the production, maintenance, and use of military hardware.

(thin-film deposition and growth) and subtractive (etch) processes. This has evolved into an enormously sophisticated enterprise, which is proven for the low-cost, high-yield manufacture of extremely complex (~100 million transistors) and reliable circuits. However, limitations are on the horizon. One is the difficulty of extending today's optically based lithographic techniques to the nanoscale, which is much smaller than ultraviolet optical wavelengths. Another is the limited number of materials used in ICs. The nanotechnology community is investigating many disparate technologies based on many different materials, but it is far from evident that the different processing requirements of these technologies can be reconciled.

Self-assembly is a radically different approach to fabrication at the nanoscale. It takes advantage of molecular and intermolecular forces to define atomic, nanoscale, and macroscale structures. Self-assembly depends on appropriate direction and control being exerted at all stages of the process by preprogramming of the subunits or building blocks such that the required recognition elements for self-assembly are contained in the subunits. Crystal growth is an example of self-assembly with exquisite long-range order. Living species are proof that complex three-dimensional structures with interacting functionality are possible. Integration of the top-down (lithography and pattern transfer) and the bottom-up (self-assembly) approaches offers an attractive approach to bridging the current gaps between these paradigms.

The incompatible materials issue may be addressed by individualized optimization of different devices and subsystems, followed by an assembly process akin to the automotive assembly line but at a vastly smaller scale. Here again, top-down (pick-and-place) and bottom-up, self-assembly inspired (DNA-assisted) approaches are among the many being investigated.

Lithography and Pattern Transfer

The microelectronics, computer, and information revolutions can trace their success to several technological roots. The integration of transistors into functional blocks—which are further integrated to form microprocessors, memories, and other ICs—is one of the main reasons for the ever-increasing functionality. Decreasing linewidths, currently 130 nanometers, enable placing more transistors on a chip. Increasing wafer sizes—now up to a 300-millimeter diameter, with projections to 400 millimeters—allow more chips to be produced simultaneously. Mass production of integrated circuits by batch fabrication with high yields has led to declining cost per function and the remarkable reliability of microelectronic devices. Not having to assemble individual parts before packaging chips has also been a major factor in the high yields and low costs of microelectronics. A great triumph of microelectronics has been the high-yield manufacturing of reliable 100-million-part assemblies. With nanotechnology, reliability of even more complex assemblages is possible.

There are fundamentally only two kinds of things that have to be done to produce integrated circuits—a pattern must be made and it must be transferred into the work piece by either deposition or removal of materials.[1] The pattern definition is done by lithography. Pattern transfer involves any of a number of processes for adding materials to a wafer, such as ion implantation, or removing materials, such as plasma etching. Ancillary techniques are also employed to ensure that the pattern production and transfer techniques work properly. Planarization of a partially processed wafer by chemical-mechanical polishing is an example.

Currently, lithographic exposures are done between 24 and 30 times or so during production of a complex IC. Each exposure requires multiple processing steps, including spinning on, prebaking, exposing, postbaking, developing the resist, pattern transfer, and removing the residual resist. Indeed, there may be about 200 processing steps for a modern integrated circuit. Today the mask sets for the most sophisticated circuit can cost about $1 million. The systems that align a given mask level to structures on a wafer and make the exposures cost about $10 million. Aligners are projected to cost close to $30 million in 5 years.

Lithography enables production in many technologies besides integrated circuits. Figure 4-2 shows several of the classes of structures and devices that require lithography to produce a pattern on a substrate. The highest resolution lithography is generally required only for cutting-edge integrated circuits. However, techniques developed in the IC industry clearly have applications to many other fields. In the top row of Figure 4-2 is an etched integrated circuit with its copper interconnects, a flexible printed circuit board, and solder bumps. The center row shows a deep-etched silicon structure, a photonic material, and a microfluidic device. In the bottom row are a gene chip, densely packed nanocrystals, and guided cell growth.

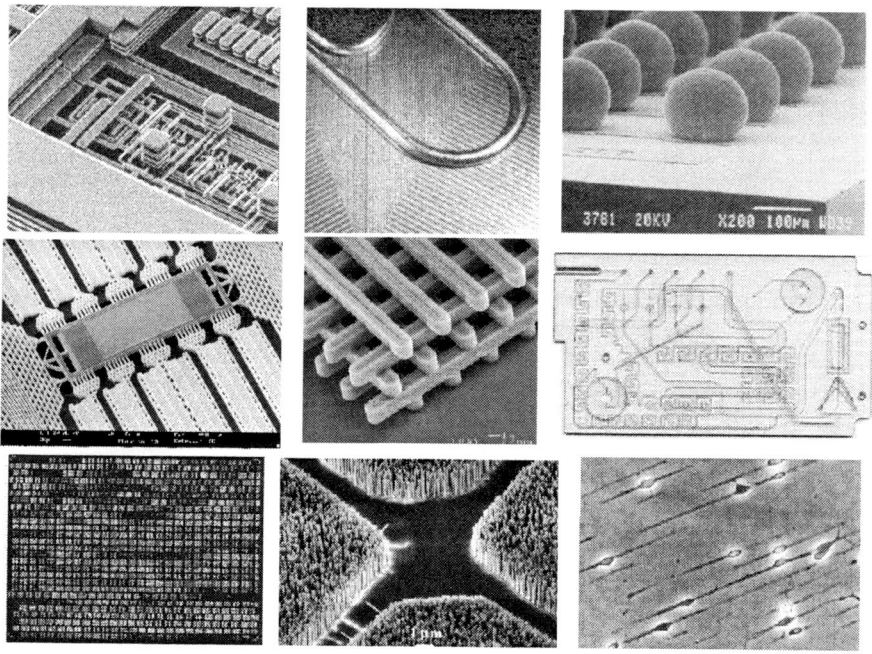

FIGURE 4-2 Lithography examples. SOURCE: Nagel, D.J. 2002. Technologies for micrometer and nanometer pattern and material transfer. Direct-Write Technologies for Rapid Prototyping Applications, A. Pique and D.B. Chrisey, eds. New York, N.Y.: Academic Press. © 2002, Elsevier Science (USA), reproduced with permission of the publisher.

Currently, most aligners use 248-nanometer radiation from KrF lasers to expose photoresist. The switch to ArF lasers, which have a wavelength of 193 nanometers, has already begun for the production of chips with 100-nanometer linewidths. In a few years, the ITRS indicates that it will be necessary either to continue this migration to a 157-nanometer wavelength advanced optical lithography tool or to switch to a next-generation lithography approach. Both alternatives are being intensively investigated and are likely to coexist for some time.

The leading contender in the United States for the next lithography technology is extreme ultraviolet (EUV) radiation with a wavelength of 13 nanometers.[2] EUV lithography requires the use of plasma sources produced by high-power laser irradiation of atomic xenon, all reflective optics (including the mask), and—as usual for a different wavelength—new photoresists. Numerous technical barriers must be overcome for EUV lithography to be ready for the production of commodity ICs.

In Japan, there is still strong interest in x-ray lithography using 1-nanometer wavelength radiation.[3] In this case, the "light" source would be either plasmas or high-energy electrons orbiting in an evacuated toroid, so-called synchrotron radiation. A fundamental challenge with x-ray lithography is the mask, which must have features the same size as the pattern to be impressed into the photoresist. That is, x-ray lithography is a one-to-one technology and not a projection reduction technique, because optical elements that focus 1-nanometer radiation are not available.

Techniques to further extend optics using alternative exposure schemes and various nonlinear processes in the lithography process are also under investigation. It is far from clear which lithography technique will prove to be the industrial workhorse for ICs with linewidths below 100 nanometers. It is possible that electron projection or direct-write technologies will play some role in future chip production. For large-volume commercial applications, economics will be a major driver. Evolution of an incumbent technology, e.g., optical lithography, almost always wins until fundamental limits are reached. For optics the fundamental limit is related to, but is not the same as, the optical wavelength. In a single exposure the limit is on the highest spatial frequency, or how close together two features can be located, not on how small each feature can be. This limit is approximately one-fourth of the wavelength, or 50 nanometers for today's ArF-laser-based lithographic tools. Immersion techniques allow a reduction by a factor of about 1.5, the refractive index of the immersion fluid. Multiple exposures taking advantage of inherent nonlinearities in the photoresist and subsequent processing stages allow further extensions by factors of one-half, one-third, etc. The manufacturing limits for these optics extensions are associated with process latitude and yield rather than with fundamental physical limitations.

It is clear that lithography for the volume production of microelectronics is already a nanotechnology and will become increasingly so in the next decade. Moreover, it will be a long time before any of the methods of modern nanotechnology, such as the growth and use of carbon nanotubes, rival lithography in commercial volume.

Beyond the methods that are in use or in contention for mass production of integrated circuits, a number of lithographic methods have been developed in the past decade. Not all of them offer the nanometer resolution that will be needed for IC production in coming years. However, they have been or might prove to be of use for making some MEMS and other structures and devices. These methods are briefly reviewed before discussing pattern transfer methods.

A three-stage process, developed in Germany, involves sequential use of lithography, electrodeposition, and molding (Lithographie, Galvanoformung, und Abformung) (LIGA). In contrast to the lithography technologies surveyed so far, which use resists with thickness from nanometers to micrometers, LIGA employs resists with thickness from micrometers to more than millimeters. The steps in the LIGA process are shown schematically in Figure 4-3.[4]

The next lithography method is both the most recently discovered and the most unconventional compared with commercial techniques. It is called lithographically induced self-construction (LISC),[5] and it has a variant termed litho-

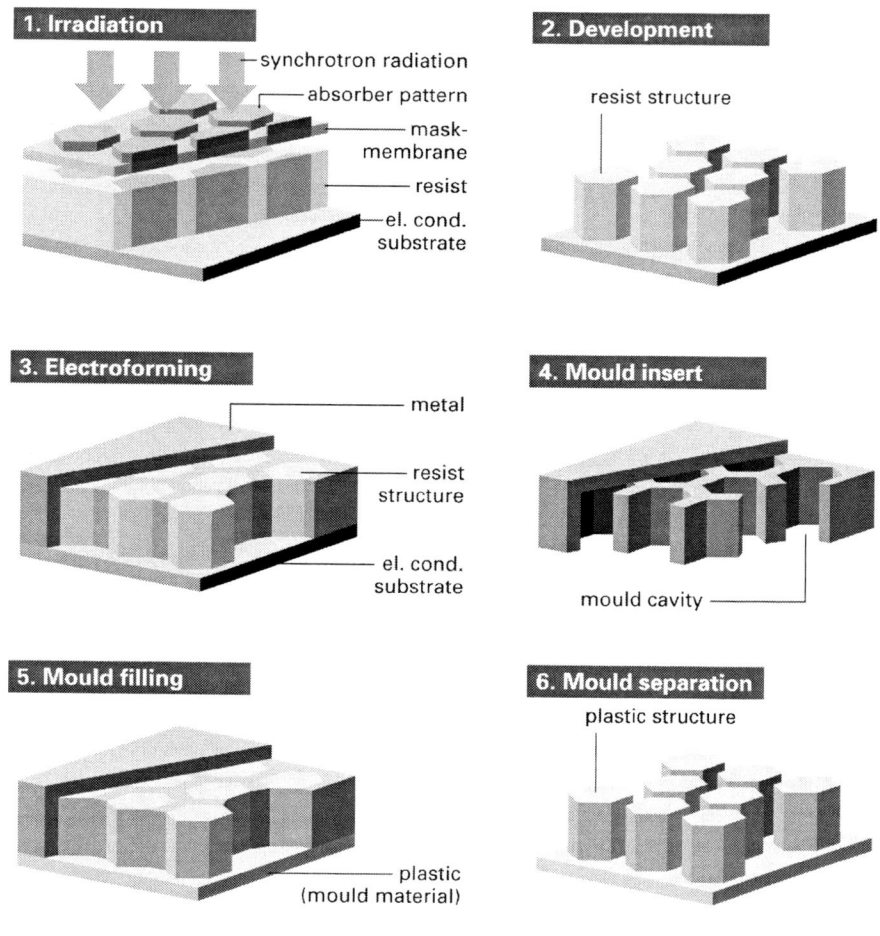

FIGURE 4-3 The sequential steps in LIGA. In 1 and 3, el. cond. means electrically conductive. SOURCE: Institut für Mikrotechnik Mainz GmbH. 2001. LIGA Technology. Available online at <http://www.imm-mainz.de/> [August 12, 2002].

graphically induced self-assembly (LISA).[6] In both cases, a patterned mask coated with a surfactant is placed in close proximity to a substrate covered with a thin thermoplastic polymer (Figure 4-4). The polymer may be chemically identical to a photoresist—for example, polymethyl methacrylate (PMMA)—but it does not function as a normal photoresist. There is no radiative transfer between the mask and polymer, and the chemistry of the polymer is not modified during pattern transfer. Rather, when the PMMA is heated to 170°C (which is above the softening point), electrostatic forces cause it to move laterally into shapes mediated by the nearby mask. The LISC and LISA techniques are both hybrid contact and

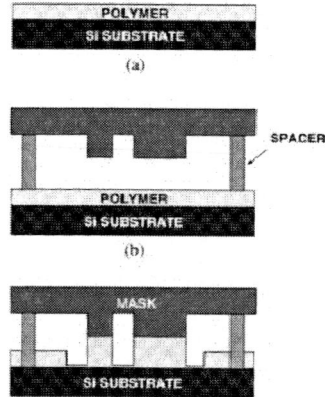

FIGURE 4-4 Schematic of the structures used in LISC. SOURCE: Reprinted with permission from Chou, S.Y., L. Zhuang, and L. Guo. 1999. Lithographically induced self-construction of polymer microstructures for resistless patterning. Applied Physics Letters 75(7): 1004–1006. © 1999, American Institute of Physics.

proximity methods. In LISC, the resulting pattern is determined only by the pattern used. In LISA, the pattern that is produced has an overall shape set by the mask but fine details within that shape that arise spontaneously.

Embossing is one of the old technologies that has been extended to the micrometer scale in recent years. Hot embossing can be employed to produce micrometer and even nanoscale features and structures. Patterns with structures about 10 μm wide and deep can be embossed into PMMA.[7] Recently, a modified wafer bonding system was used to emboss structures as fine as 400 nanometers across an entire 10-cm-diameter wafer.[8]

A variant of embossing, called nanoimprint lithography (NIL), involves impressing a mold onto the surface of a photoresist-covered substrate.[9] In this case, the pattern is transferred to the resist, commonly PMMA, by mechanical rather than chemical action. Subsequent processing of the resist to open the thinned regions to the substrate permits conventional uses of the resist for deposition onto or etching into the substrate. The molds for NIL can be prepared by a wide variety of the normal and developmental lithographic processes. For example, e-beam lithography has been used to make a mold with 10-nm-diameter pillars on a 40-nm pitch that were then imprinted into PMMA.[10]

A technique called step and flash imprint lithography essentially embosses a layer of liquid on a surface that is then turned into a solid using a photochemical process.[11] This technique avoids the elevated temperatures and pressures ordinarily required for embossing. The wafer is first coated with a transfer layer of solid organic material. Then a glass template with the desired pattern is placed near the coated wafer. The template can be micromachined by a variety of methods. A low-viscosity liquid (a photopolymerizable, organosilicon etch-barrier material) is dispensed between the template and transfer layers on the wafer, and the layers are then brought into contact. After UV exposure to solidify the etch barrier and make it adhere to the transfer layer, the template is removed. A plasma etch transfers the pattern from the now solid etch barrier into the transfer layer, and then the etch barrier is removed. This leaves the pattern in the transfer layer on the wafer surface ready for further processing steps. Step and flash imprint lithography has produced 60-nanometer features.

Rubber stamps have been used for centuries. In recent years, Whitesides and his group have extended stamping to replicate patterns with features finer than micrometers by the use of polydimethylsiloxane (PDMS) and other elastomers. They term the technique "soft lithography" because of the compliant character of the stamp.[12] One of the ordinary lithography methods is used to pattern a thin film on silicon or some other substrate to make a mold. After the surface of the mold has been etched and silanized, the liquid PDMS precursor is cast over the pattern and polymerized by cross-linking. Then, the elastomer is peeled off the mold and placed on a substrate for handling. Wetting of the PDMS stamp with various liquids and suspensions is done before the stamping. In general, the depth of the pattern is in the range of 0.2 to 20 µm, with the maximum limited by the stability of the PDMS structure. The width and spacing of the contact regions are between 0.5 and 200 µm, with the separation limited by the tendency of the region between contacts to bulge toward the substrate being stamped. Soft lithography can be employed in a rolling manner if the PDMS stamp is attached to a cylinder.

The methods for pattern production and transfer that are used in the manufacture of microelectronics IC production involve a very limited number of materials and associated processes. The materials used for chips now number only about 10. Copper interconnects, high-dielectric-constant (k) materials for gate insulators and low-k materials for separating interconnect lines are the most recent additions to the list of IC materials. The processes used to make these materials during chip production also number about 10.

IC technologies are certainly fundamental to the production of microscale mechanics, optics, and magnetics. However, the number of materials and processes in demonstrated and emerging MEMS and similar technologies far exceed those used for IC production, as indicated schematically in Figure 4-5. One example is the use of piezoelectric materials in MEMS for linking electronic and mechanical behavior. The substrates employed for MEMS devices include semiconductors other than silicon, notably silicon carbide, as well as ceramics, metals, and polymers.

There are numerous processes used in the manufacture of MEMS devices that play no role in IC production. The central step in the production of micromechanics on a substrate by surface micromachining is the dissolution of a sacrificial layer of material to release the mechanism. Deep etching of a substrate, called "bulk micromachining," is another important MEMS pattern transfer tech-

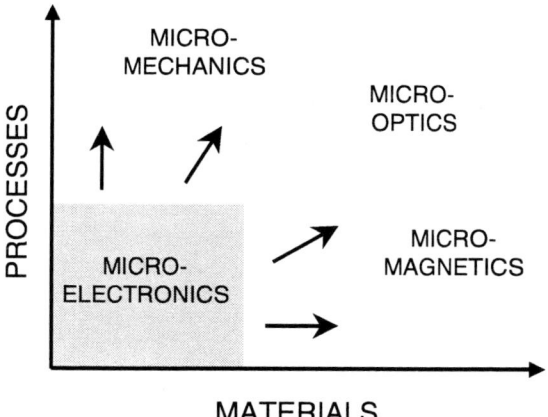

FIGURE 4-5 Integrated circuit production. The number of materials and associated processes for the production of micromechanics, -optics and -magnetics greatly exceeds the number of materials and processes used to manufacture microelectronics.

nology that has no role in IC manufacture. Such etching is done by one of two processes. The use of solvents, which attack different planes in silicon at widely varying rates, is called orientation-dependent etching. It now plays a role in the production of some MEMS products. The alternative is deep reactive-ion etching (DRIE). In this case, the plasma ambient over a silicon substrate is alternated every few seconds between etching and passivation (coating) conditions. This process, invented by the Bosch Corporation, can result in deep, narrow trenches in the substrate, the walls of which are scalloped on a very fine scale. Figure 4-6 shows examples, and it also indicates that the rate at which the Bosch process etches into the silicon is geometry dependent.[13]

The role of pattern transfer for the production of nanoscale structures and devices will probably be similar to that for the production of microscale structures and devices involving mechanics, optics, and magnetics. That is, the transfer processes used in microelectronics manufacture will remain important in many cases. The self-assembly of complex structured materials, as discussed in the next section, is playing an increasingly important role in nanoscale fabrication. New approaches using lithography to direct self-assembly are already emerging in fields as disparate as semiconductor crystal growth (discussed below) and carbon nanotube formation and are likely to play a significant role in the development of nanotechnology. The wider variety of materials that will play a role in nanotechnology will probably involve additional techniques for pattern transfer. This will almost certainly be the case for nanoscale structures that use organic and biomaterials.

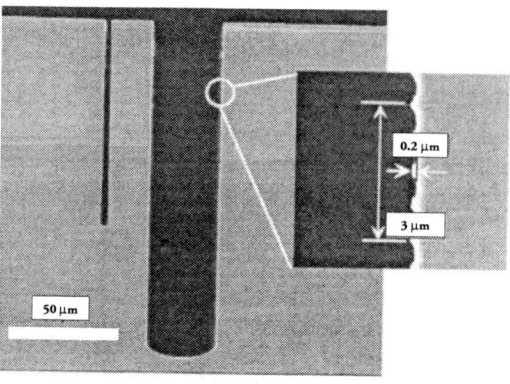

FIGURE 4-6 Cross-sectional photograph of a silicon wafer processed by deep reactive ion etching. SOURCE: S.D. Senturia. 2000. Microsystem Design, p. 70. Boston, Mass.: Kluwer Academic Publishers.

Self-Assembly

Self-assembly in nature is older than life itself. Crystalline geological structures spontaneously form highly organized patterns from molten lava. All living organisms, from the simplest single-cell species to humans, depend on some form of molecular self-assembly. Nature performs the most astonishing feats of self-assembly with an artistry and facility that we can only admire and all too often take for granted. Protein folding, nucleic acid assembly and tertiary structures, phospholipid membranes, ribosomes, microtubules, and the nucleocapsides of viruses are but some, representative examples of biological self-assembly in nature that are of critical importance to living organisms.[14]

Aside from their ability to carry out the functions of life, one of the more remarkable features of self-assembled materials is that their structure may be very complex yet demonstrate long-range order. The power and beauty of spontaneous self-assembly derives from its ability to rapidly, and with seeming ease, generate large, complex, sophisticated "supermolecules," or ensembles of molecules, from easily available building blocks with high efficiency, generally under mild conditions (at or near room temperature, at atmospheric pressure, in water or other common media). A biological self-assembly,[15,16,17] as exemplified by such structures as crystals,[18,19] surfactants,[20] micelles (nanoscale molecular aggregates),[21] colloidal suspensions in confined films or aggregated into fractal structures,[22] self-assembled monolayers (SAMs),[23] and liquid crystals,[24] involves ensembles of molecules with unique properties and function. Because it is possible to manufacture nanostructures with useful properties through self-assembly, interest has recently increased substantially in the study of interactions of molecules that form larger nanostructures of ordered material aggregates.

Nature's repertoire of information to guide self-assembly includes hydrogen bonding, π-π stacking, electrostatic and van der Waals interactions, hydrophobic-hydrophilic interaction, dipolar effects, molecular conformations, and phase boundaries, as well as the shape and size of both the final product and the constituent subunits. These effects are commonly referred to as weak interactions in contrast to the strong chemical bonds characteristic of simple molecules,[25] which were the basis of the revolution in chemical products of the last 100 years.

An important feature of all self-assembly processes is the existence of a kinetically labile, reversible equilibrium between starting materials, intermediates, and products,[26] with the final outcome under thermodynamic control (i.e., the end ensemble is the thermodynamically most stable one). As the equilibrium is reversible, the process is self-correcting; an incorrectly formed bond can disassociate and reassociate correctly. As a consequence, self-assembly processes generally engage in self-repair and are self-healing.

Self-assembly has the potential for bottom-up fabrication techniques to produce nano- to mesoscale materials and systems, that is, structures and ensembles having spatial dimensions in the range of 1 nm to 1 μm—the size of a large molecule to the size of a living cell. Nano- to microscale systems bridge the molecular and the macroscopic and display unique collective and often nonlinear behavior and properties, different from the bulk characteristics of common substances. Control of the bottom-up products is difficult at present (except for biologically produced and a few select materials) because of the currently limited understanding about how to use these forces. The current state of knowledge about self-assembly and the potential for new products may be comparable to the state of knowledge about the chemical bond at the turn of the previous century. Expanding interest in the fabrication of new materials with these self-assembly methods may give rise to an entirely new discipline.

Examples illustrating the potential of self-assembly as a process for materials production include materials such as the shell of an abalone. This composite material is formed by the careful assembly of elongated calcium carbonate crystals in layers, with the long axes of the crystals pointing in perpendicular directions in alternating layers. A protein deposit forms a strong glue to hold these crystals in place, forming the strong final product.

A wide variety of useful composite whisker structures are formed by a number of mechanisms such as condensation on specific crystal faces, selected chemical reactivity, and eutectic behavior of alloy mixtures. An example of a fascinating nanostructure formed through the forces involving chemical bonds is that of a carbon nanotube. Selected nanotubes with high conductivities have excited researchers because they might one day be used as conductors and circuits in nanoelectronics (see Box 3-1).

Block copolymers form amazingly regular patterns of plastic material with repetition distances on the order of nanometers to micrometers. These plastics interact with light and are used in devices such as optical band gap structures for optical communications. Block copolymers can be etched to form porous arrays of regularly spaced voids. Porous materials may be formed that have selected void dimensions for specialized catalytic activity. Likewise, dendrimers and dendrimer assemblies,[27] as a result of their globular shape, their layered architecture, their well-controlled size, and their unique optical and electronic properties, may find uses in micro- and nanoelectronics and other applications.

When fully developed, bottom-up fabrication using self-assembly and related methodologies will be able to produce ordered, precisely controlled nanomaterials and nanodevices not achievable by current methods such as lithography and other top-down techniques.[28] Recent examples of interest include novel protein-based materials such as artificially modified spider dragline silk, mussel byssus thread, elastin, and other higher-ordered aggregates derived by protein engineering.[29] Other directions of self-assembly research include genetic methods of polymer synthesis;[30] kinesis-powered microdevices (molecular motors);[31] nano-optics in the biological world;[32] molecule-based magnets;[33,34] biomedical applications;[35] and molecular-based materials in electronic devices, light-emitting diodes, quantum dot lasers, and photovoltaic devices.[36]

As a field of endeavor with widespread applicability, self-assembly is just beginning a new era for the production of new materials.

Integration of Traditional Lithographic and Self-Assembly Patterning Approaches

One of the grand challenges of nanoscience is precise positioning, atom by atom. Historically there have been two approaches to this challenge: top-down lithography and bottom-up self-assembly. Lithography has made enormous strides during the development of the integrated circuit. As discussed above, present lithographic limits are in the 100-nanometer range. New techniques are being developed that will reduce this scale by approximately an order of magnitude, still far from a typical molecular interatomic distance of approximately 0.5 nanometers.

From the other direction, using self-assembly techniques, chemists routinely make molecules, literally by the boxcar full, with precise distances between atoms. Crystals are beautiful examples of precise self-assembly with long-range order. On a 300-mm-diameter silicon wafer, if you know the position and orientation of a single unit cell, you know—to within variations due to temperature, stress, impurities, and thermal motion—the precise positions of the atoms on the other side of the wafer, or at the other end of the boule. Recent explorations of molecular self-assembly have sought to provide transverse dimensions on the mesoscopic nanometer scale. As a general—although not inviolate—rule, these attempts have led to very good local ordering (e.g., nearest neighbors) and comparatively poorer long-range order (on macroscopic dimensions). This is inherent in techniques that provide multiple nucleation sites, for example on a flat, non-crystalline surface. Multiple crystal domains nucleate as dictated by the thermodynamics of the process, and grain boundaries result as the individual crystallites grow and merge.

One active direction of research is to meld these two techniques and use each to its best advantage. Top-down lithography can provide a very high degree of long-range order on from the centimeter scale down, with features that now extend down to about 100 nanometers. Molecular self-assembly works well below the 10-nanometer scale. These regimes are close enough that they will merge and offer the possibility of complete control from the macroscopic to the atomistic. Nanotechnology will likely demand control across all of these spatial scales.

In the information technology domain, a further motivation for addressing all of these scales is that the new nanoscience functional elements—e.g., the equivalent of the transistor and its interconnections—are most likely to be used in combination with our existing silicon-based electronics infrastructure. Thus, we will already have the top-down processing in place.

In the vertical dimension, there are many examples of using self-assembly as a very precise ruler. The exceptional precision of today's epitaxial growth techniques was discussed in connection with quantum wells in the first section of Chapter 3, "Information Technology." As a result of this precision, it is literally possible to count atoms as an epitaxial layer is being grown and stop when that layer is any integral number of atomic layers thick. In some cases, one can even transmute the layers after they are grown, as is done in the case of defining 1/4 distributed Bragg reflection mirror stacks for vertical cavity lasers by selective oxidation of high aluminum content $Al_xGa_{1-x}As$ layers to aluminum oxide.[37] Recent reports have discussed using similar techniques in the transverse dimension by building up a set of molecular multilayers at the edges of masks defined by top-down lithography.[38] These approaches are still in their infancy, and much has to be done to mature them into useful applications. They represent one of many conceivable ways of combining the top-down and bottom-up approaches to define nanoscale features.

Semiconductor Epitaxial Growth—An Example of the Integration of Bottom-Up and Top-Down

Semiconductor growth is a rich arena for innovations in patterning. The recent interest in three-dimensional confinement (quantum dots) and in photonic crystals (see section on optical devices in Chapter 3) has led to a spurt of activity. Traditional semiconductor epitaxial growth is carried out on large-area substrates, up to 300 mm in the case of silicon. While there is a rich tradition of growth on patterned surfaces, the scale of these surfaces has traditionally been several micrometers or larger, used primarily to define isolated growth areas.

Two notable early exceptions were (1) the use of slightly misaligned silicon substrates with a well-defined series of atomic steps to reduce antiphase disorder (both Ga-rich and As-rich <100> faces of GaAs nucleate on a Si <100> face) in the growth of GaAs on silicon,[39] and (2) the use of etched V-groove structures as substrates for the growth of quantum wires just at the apex of the V-groove.[40] Neither of these early attempts at nanopatterned growth proved successful for growing device-quality material because of issues associated with remaining defects. A successful microscale patterned growth approach has been the use of the dependence of growth rate on area and on crystal face in the case of deeply etched structures to achieve multiple wavelength devices in the same growth.[41] This is particularly important for network wavelength division multiplexing (WDM) applications, where a range of wavelengths is required. Having all of the sources (and possibly matched detectors) on the same substrate is an important packaging advantage.

With the increasing ability to fabricate nanostructured substrates, there is increasing interest in epitaxial growth at these dimensions, an area that is largely unexplored. One specific direction is the use of nanostructured growth for heterogeneous material systems with significant lattice and thermal mismatches between the film and the substrate.[42] Much of this interest has been spurred by the device applications of wide bandgap GaN and related materials for which there is as yet no suitable bulk substrate. Applications are both commercial (visible LEDs and lasers, high-frequency and high-power electronics) and military (solar-blind detectors, UV sources for chemical and biological agent detection, high-frequency and high-power electronics). Microscale structured substrates have been successfully developed based on epitaxial lateral overgrowth (ELO), where the film is allowed to nucleate in only small areas and grows laterally over the masked areas with many fewer defects. Nanostructured materials, including self-assembled porous substrates, block copolymer films, and lithographically defined patterns, are being investigated. There appear to be several advantages to the nanoscale in that nanoscale seeds reducing the epitaxial film strain can accommodate some of the lattice mismatch. Also, the local free surface associated with the nanoscale appears to modify the character of the remaining defects, in some cases resulting in a confinement of these defects to the near interface region. Much more work is needed to determine the utility of these results for actual device materials, but the initial results are promising.

The direct fabrication of quantum-confined semiconductors is another important direction. As noted in Chapter 3, attempts to define quantum dots by subtractive processes (etching) have been stymied by significant defect densities, which lead to rapid nonradiative recombination. Growth processes wherein a nanoscale mask is first defined and the material is then grown have shown some promise, but still exhibit poorer photoluminescence than fully self-assembled materials and have not yet produced device-quality material. Nanoscale patterned growth remains a promising active area of research with potential for important advances. An alternative direction that shows nascent promise is the adaptation of some of the techniques used to grow carbon nanotubes to more traditional semiconductor materials—both Si and III-V materials.[43,44] These techniques have now been extended to junctions, heterostructures, and superlattices demonstrating a wide array of functionality, including light emission and electronic nonlinearity.[45]

Integration of Nanodevices with Mainstream Silicon Technology

Modern electronics is based on silicon technology. As noted elsewhere in this report, progress in silicon technology and manufacturing has been steady for perhaps 50 years and is projected out at least 15 years into the future and certainly well into the nanoscale regime. This progress and the utility of electronic semiconductor products caused semiconductor manufacturing to become one of the largest industries in the world and certainly one of the most sophisticated and creative.

The latest release of the ITRS[46] forecasts significant difficulties over the 2002-2017 time frame in various perhaps fundamental aspects of the scaling that has held for so long. As detailed in Chapter 3, this has led to a host of efforts to discover the successor to silicon. The ultimate utility of any of these pretenders to the silicon crown is uncertain, and predictions of test-tube microprocessors seem premature. Some numbers make clear the issues: today's most advanced memory chips contain on the order of 10^9 transistors, interconnected in a complex pattern involving a hierarchy with both local and global character. Not only the internal functioning, but also the input/output protocol, is precisely defined and matched to the requirements of the microprocessor interface buffer. Together these elements perform an elaborate symphony to provide function. And despite the very large cost of the equipment and manufacturing infrastructure, circuits are produced so quickly and reliably that the cost of computing power, measured in computations per second to the end user, decreased by over nine orders of magnitude as the technology progressed from vacuum tubes to discrete transistors and to today's advanced microprocessors over the past 50 years.[47]

In contrast, each of the contenders for silicon's crown is today largely at the stage of working out what its "transistor" is going to be. Even if this "transistor" turns out to have exceptional performance (which so far remains possible but not proven), there is still a very long way to go to integrate the "transistor" with everything else and produce a computer.

This is not to suggest that research in alternatives to silicon electronics is not warranted. But, it may be that these new technologies will first find use as complements to silicon, not as immediate replacements. We already know how to do so much with silicon it seems unlikely that a new technology will spring up overnight sufficiently developed and robust to supplant not only the transistor but all the rest of the system.

ASSEMBLY

Webster defines the term "assembly" as the "fitting together of manufactured parts into a complete machine"—for example, interfacing disparate components into a functional hybrid unit (think of what an automotive assembly line does). At the beginning of the 21st century, assembly is moving to smaller and smaller dimensions and to very-high-technology processes and materials. This section will discuss two new types of micro- and nanoassembly to illustrate the potential richness of these new, very small assembly methods: directed assembly (e.g., programmed pick-and-place) and fluidic self-assembly relying on either shape keying or DNA-labeling. In many ways the origins of these types are akin to the top-down and bottom-up patterning technologies discussed above.

Directed Assembly

Directed assembly is distinguished from self-assembly, where quasi-equilibrium environments are used to arrange things. (An example of self-assembly is the microfluidics, discussed in the next section.) Directed assembly is programmable assembly, where parts are moved mechanically and placed precisely where they are intended to go.

The impressive gains by the IC industry have been accomplished by parallel processing and monolithic integration. However, monolithic integration has limits, especially as the materials repertoire is increased beyond traditional electronic materials. The lack of a monolithic integration process extending across a wide range of functional materials demands that discrete parts be assembled into a whole. Unfortunately, directed assembly technology has lagged seriously behind monolithic integration in miniaturization. Consider that while an individual transistor is significantly smaller than a micron on a side, the smallest parts that are routinely assembled by today's automated assembly tools are approximately 1 mm on a side. Not coincidentally, this is essentially the smallest part a human can reasonably manipulate in manual assembly operations.

Robotics and automation experts are starting to build assembly systems to handle parts smaller than 1 mm. However, these systems and their unit operations get more expensive as the parts and required precision of assembly decrease in size. This is true principally because the parts for the automation tools are produced by serial processing (conventional machine shops) and the automation assembly process itself is serial.

Progress in directed assembly needs to take advantage of the high-precision, parallel processing that the IC industry has developed. This strategy will provide a clear path to scaling down components and microsystems, which may be more cost effective than either manual or conventional automation assembly. An example of this type of process is the parallel assembly process being developed by Zyvex, Inc. This company is using silicon MEMS as an integral part of the assembly system: as a method of arranging parts for assembly, as the fastening mechanism, and as a method for fine positioning adjustments.

Several factors argue for using MEMS to manipulate the parts being assembled. The starting point is that submicrometer dimensional control is readily available through MEMS but is very difficult and expensive with conventional machining. The excellent mechanical properties of silicon are an advantage. Another factor is the well-developed infrastructure that already exists for silicon materials and processing. If there were any questions about the fundamental reliability of MEMS devices, they should have been put to rest with the success of Texas Instrument's (TI's) digital light projector chips (see the last section of this chapter for the history of this product), which are by any reasonable measure the most reliable mechanical devices ever created. Finally, there is a clear roadmap for downscaling MEMS that has already been formulated by the IC industry. Direct assembly is aimed at overcoming one of the principal limitations of MEMS processing, its generation of essentially two-dimensional parts. Assembling MEMS parts will enable multiple degrees of freedom of motion.

The key points of this assembly strategy include a movable tether approach that allows parts to be arranged on silicon wafers and constrained in a known position until they are captured. To achieve high throughput, the parts and finished goods are arranged in a periodic array to allow parallel assembly. The parts consist of modular MEMS carriers that have either integrated MEMS devices or preattached component parts. The carriers include silicon snap connectors that make mechanical and, if needed, electrical connections with the assembly substrate. An array of MEMS grippers is attached to the macroscale robotics used to capture parts from the parts array and carry and attach them to the assembly arrays in a highly parallel manner. Figure 4-7 shows an assembled MEMS gripper device, called a rotapod, used to capture and position parts. The rotapod is a MEMS device assembled from two separate MEMS parts using snap connectors. It has a gripper mechanism and is capable of rotating along two axes. This part will be able to capture parts and rotate them into position for assembly, thus breaking out of the dominant two-dimensional MEMS assembly paradigm.

The grippers and snap connectors are designed with self-centering mechanisms that allow significant tolerance in the capture and attachment process. This self-centering capability effectively discretizes the assembly process, allowing high-precision assembly with relatively inexpensive robotics. The silicon snap connectors place components more accurately than the robotics.

The main advantage of such a MEMS-based manufacturing technique over current practices is parallel processing, analogous to the advantage that IC processing enjoys over discrete electronics. Parallel processing should drive down the cost of microsystem assembly.

This assembly manufacturing technology may find its first major application and a large market share in manufacturing fiber-optic communication components. This will allow demonstration of the technology at larger size scales and

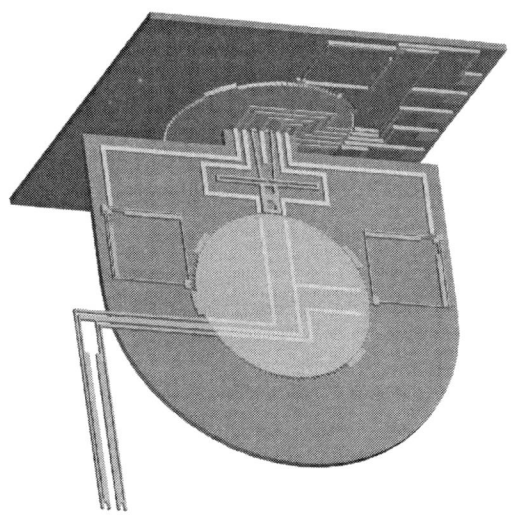

FIGURE 4-7 Rotapod MEMS device. SOURCE: Randall, J. 2001. MEMS/Nano Technology for Assembly of Systems. Briefing by John Randall, Vice President for Research and Chief Technology Officer, Zyvex, to the Committee on Implications of Emerging Micro and Nano Technology, DoubleTree Hotel, Albuquerque, N. Mex., November 8.

will pave the way for assembly of parts smaller than can be handled manually or with conventional automation equipment. Further, once a low-cost assembly manufacturing process is available, many systems being attempted with monolithic integration approaches will find much more success with separate optimized processes for different components that can then be assembled into complex functional units.

DNA-Assisted Assembly

Future photolithographic processes are expected to provide nanoscale metal oxide semiconductor transistors and other semiconductor-type devices. But truly monolithic, heterogeneous integration of photonic, microelectronic, MEMS, and microfluidic components into one chip is both difficult and expensive. The difficulty arises from the incompatibilities of the functional materials and of the various fabrication processes required to optimize each type of device. In addition, while many new molecular and nanoscale components are under investigation (Chapter 3), their homogeneous and heterogeneous integration into higher-order two- and three-dimensional structures and devices will not be straightforward with standard fabrication and lithographic processes.

As a result of these limitations, hybrid heterogeneous integration methods are being extensively investigated, for example, flip-chip bonding and wafer bonding followed by substrate removal. However, these hybrid integration techniques are limited to handling chips, not individual devices. This limitation arises from the serial and macromechanical means used to perform pick-and-place. As the device sizes shrink and as the functionality of heterogeneous integration (bringing many devices of different origins together on a chip) gains importance, high-throughput pick-and-place techniques capable of handling many nanoscale devices will become critical.

A future economically viable heterogeneous integration technique needs to provide for the rapid and parallel pick-and-place of nanoscale individual devices to desired locations on a host substrate with required accuracies and yields. Typically, biological systems use self-assembly, self-organization, and self-replication principles to achieve heterogeneous integration at nanoscale dimensions. While self-assembly and self-organization based on biological models (DNA, proteins, etc.) will be extremely useful, our present understanding of these models is naïve and inaccurate.

However, even with these limitations, some of the principles can be adapted and used for heterogeneous integration. For example, fluidic self-assembly techniques have been used to populate a wafer with an array of silicon chips.[48] In this case, devices are fabricated as specific geometrical shapes that can then slot into similarly shaped holes etched into the host substrate. The circuits are combined with a liquid and spilled out over the host, hopefully finding the right slot and sticking by van der Waals forces. Fluidic assembly methods have an important advantage over competing systems such as mechanical pick-and-place: devices are deposited in parallel, not one at a time, and many different-shaped devices can find their new location in a single step. One of the shortcomings of this approach is that since devices must be shaped, very small devices cannot be handled.

The remarkable recognition properties of deoxyribonucleic acid (DNA)-like molecules can be harnessed to overcome this limitation. Pick-and-place of many types of micro- to nanoscale devices in fluids with high throughput has been demonstrated.[49,50]

DNA-based pick-and-place techniques rely on several principles to achieve heterogeneous integration:

- DNA molecules are charged and therefore can be electrochemically transported.
- DNA molecules can be sequenced and therefore coded in large quantities.
- These sequences (codes) can be used for complementary DNA strands that recognize each other and hybridize, forming strong bonds.
- DNA strands can be attached to different devices by well-known chemical processes and therefore can act as labels for these devices.
- Because DNA strands are very small, they can be used as labels for micro- to nanoscale devices.

The technique involves taking polymers made with complementary DNA strands and using them as a selective glue. These strands consist of CGAT bases (cytosine, guanine, adenine, and thymine), which will only bind to each other in specific pairs: C with G and A with T. Consequently, an ATTTGC strand will bind very strongly to its complement (TAAACG) but not to any other strand. These materials can be coated onto the bottom of specific devices and their complements patterned onto host substrates (see Figure 4-8). When a device coated with a given strand type encounters a host area coated with the complementary strand type, it attaches via the hybridization of the complementary DNA strands (see Figure 4-9). Because these strands bond to form the well-known double-helix structure, the devices remain fixed in position until electrical contacts can be formed in a subsequent process step.

It is still important to actively help the devices to find their new homes rather than leaving things to chance in order to accelerate the pick-and-place process and increase its yield. One method of doing this is to create electric fields around the host "landing sites," thus attracting devices to a particular spot using the natural negative charge of DNA.[51,52] Another is to improve alignment of the devices. This is performed once the landing sites are occupied by applying an alternating current field around the landing site. This alternating current field anneals the hybridization process by maximizing the number of hybridized strands, thus finely aligning the devices on the host substrate. Finally, a critical aspect for many applications is achieving ohmic electrical contacts between the

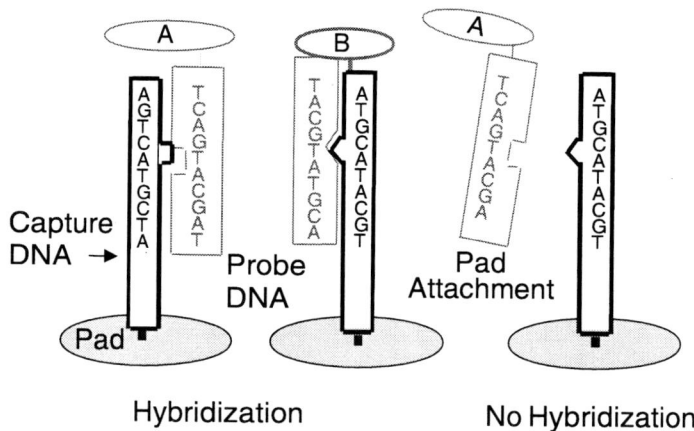

FIGURE 4-8 Principle of DNA-assisted pick and place. SOURCE: Esener, S. 2001. Emerging Applications of Micro and Nanophotonic Components and Their Interface with Biology. Briefing by Sadik Esener, Electrical and Computer Engineering Department, University of California San Diego, to the Committee on Implications of Emerging Micro and Nano Technology, National Academy of Sciences, Irvine, Calif., December 19.

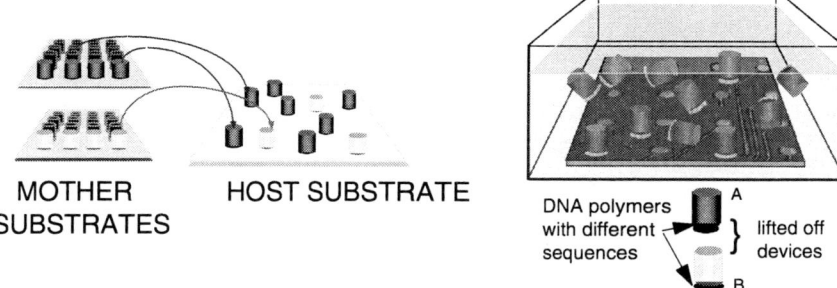

FIGURE 4-9 DNA-assisted microassembly. SOURCE: Esener, S. 2001. Emerging Applications of Micro and Nanophotonic Components and Their Interface with Biology. Briefing by Sadik Esener, Electrical and Computer Engineering Department, University of California San Diego, to the Committee on Implications of Emerging Micro and Nano Technology, National Academy of Sciences, Irvine, Calif., December 19.

devices and the host using small solder balls under the devices. After the hybridization process, the substrate is removed from the fluid and dried and the system is heat treated. Contacts are then formed in a manner similar to flip-chip bonding. Remaining DNA strands do not harm the quality of the ohmic contact. Although solder balls have been used in the initial demonstrations, many of the existing techniques for forming electrical contacts can in principle be used with DNA-assisted assembly.

This DNA-assisted assembly method is best suited to very small devices, on the order of 100 micrometers or less (all the way down to tens of nanometers), that can flow freely and find the correct position without blocking other devices.

In addition to DNA-assisted assembly, several other chemical and biological molecules can be used advantageously for device-forming pick-and-place as well as for micro- and nanoactuation. These techniques include, for example, the use of patterned hydrophobic films on hydrophilic substrates to form devices such as microfluidics and microlenses[53,54] (see Figure 4-10) and proteins for transport and actuation functions.

PACKAGING

The packaging of integrated circuits involves many complex but well developed technologies.[55,56] The packaging field has evolved in response to two major influences—cost and the need to put chips ever closer together on printed circuit boards. Dual-inline packages first gave way to surface mount packages, and these are now being superseded by chip-scale packages. Flip-chip attachment of ICs to boards essentially mates the chip to the package. Hermetic sealing of ICs from the outside environment has been a standard requirement. With the exception of

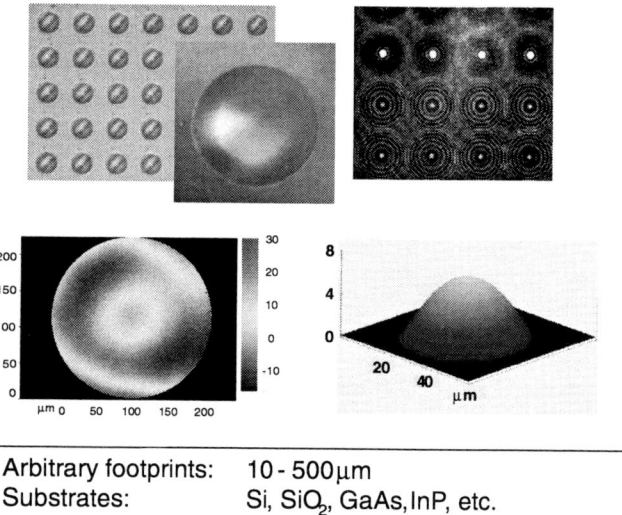

• Arbitrary footprints:	10 - 500 μm
• Substrates:	Si, SiO$_2$, GaAs, InP, etc.
• Fill-factors:	Up to 90%
• f#s:	Plano-convex f/1.38, bi-convex 1.2
• Uniform arrays:	Δf/f ~ ± 5.9%
• Surface quality:	Diffraction limited (± 300 nm over middle 9/10)
• Reproducibility:	Δf/f ~ 3.5%
• Stability:	Stable 1 year at room temperature

FIGURE 4-10 Lenslet array fabricated using hydrophobic/hydrophilic selectivity. SOURCE: Esener, S. 2001. Emerging Applications of Micro and Nanophotonic Components and Their Interface with Biology. Briefing by Sadik Esener, Electrical and Computer Engineering Department, University of California San Diego, to the Committee on Implications of Emerging Micro and Nano Technology, National Academy of Sciences, Irvine, Calif., December 19.

the windows on erasable, programmable read-only memories (EPROMs), IC packages have generally been opaque.

The packaging of MEMS devices is generally far more demanding than the packaging of ICs. The cost of packaging commonly represents 50 to 90 percent of the cost of a MEMS device. Many MEMS are in sealed packages. There are some cases in which ordinary microelectronic packages can be employed for MEMS. The primary instance is the packaging of inertial sensors, including both microaccelerometers and angular rate sensors. However, even MEMS devices in sealed packages often cannot use the same packages that have been developed for the IC industry. In fact, many MEMS have sealed packages that are significantly more complicated than standard electronic packages. Optical MEMS, which must have windows, are the primary example. In the case of the digital mirror device from Texas Instruments, the window is large and its optical properties are demanding. MEMS switches for all-optical networks also have to have optical access to the moving device, either through windows or via optical fibers that penetrate the package, sometimes in large numbers. Other MEMS in sealed packages require a vacuum environment. Microresonators must have low losses to perform properly, so air must be excluded. Uncooled infrared sensor arrays, consisting of pixels thermally isolated from the substrate, also require vacuum packaging.

There are many MEMS that cannot be used in sealed packages. Pressure sensors are a commercially important example. They are generally sealed from the ambient atmosphere by an elastomer, which transmits pressure variations but excludes humidity and dirt. MEMS strain sensors sometimes have packages that are penetrated by an element that is attached to the piece being measured. Some MEMS, like chemical vapor sensors, must be in contact with the atmosphere to function. Sensors for the analysis of water, blood, and other liquids must also have access to the exterior environment. Maintaining the sensitivity of gas and liquid analyzers over long periods is challenging.

There are some MEMS that have no distinct packages because they are located on the surface of an object or in it. The microflaps that are being developed for the control of aircraft and the microthrusters for steering aircraft or spacecraft are examples. Here also, the needed interaction between the MEMS devices and the surroundings can be problematic. Dirt and rain in the atmosphere and particles and ions in low Earth orbit all challenge the long-term use of distributed MEMS on aerospacecraft.

It is likely that the packaging of sensors, actuators, and systems based on nanotechnology will prove as varied and challenging as the packaging of MEMS devices and systems. Very small devices can be inserted into living tissue without packages. Quantum dots can be inserted into living cells for a variety of purposes. Already, their fluorescence is being used as tags for specific molecules in cells. The coating of foreign objects within living organisms by biological processes is a difficult issue that must be solved before long-term use of these technologies can be considered.

RELIABILITY AND MANUFACTURABILITY

Initially, both the yield and the reliability of nanoproducts are expected to be low. Because of the scarcity of data made available for reliability analysis, extensive burn-in testing will be needed to screen out the early failures among newly designed and manufactured products. Optimal physical conditions for reliability testing of nanoproducts must be carefully investigated until the correlation of these conditions with failure mechanisms is better understood. It is also critical to use engineering experience to predict and to improve the future batch manufacturing processors using Bayesian statistics. The tremendous experience gained in microelectronics manufacturing will be useful for the generation of nanoproducts as well.

Bayesian analysis is an efficient way of adjusting to shorter-loop manufacturing cycles. Using information available for similar systems, nonparametric Bayesian models can be used to transfer knowledge gained from one generation of models to the next generation.[57] This method can also aid in the modification and adjustment of fabrication and processes for nanotechnology. Also, burn-in can effectively improve product reliability and enhance product yield.[58]

New Techniques for Reliability Improvement

From a manufacturing standpoint, process technologies for deep-submicron devices (<0.13 micrometer) are approaching physical limits. It is difficult to achieve high performance, high packing density, and high reliability. This manufacturing process requires a high initial investment and is very expensive to operate. Thus, cost reduction by developing new techniques or approaches becomes urgent. From a reliability point of view, accelerated life test and end-of-line failure analysis become less reliable as the chip shrinks and the devices become more complex. The simple failure analysis method of sampling the output of a manufacturing line must give way to new methods in order to better understand and control the input variables at each point in the manufacturing process.

The requirement of new techniques leads to the development of built-in reliability, wafer-level reliability (WLR), and qualified manufacturing line (QML) approaches.[59] In the WLR approach, both processes and designs affect reliability. Since the traditional reliability approaches may not support enough test time or test parts to resolve failure rates as low as 10 FITs (1 FIT = 1 failure per 10^9 device-hours), considered the goal for failure rate, approaches for improving reliability and yield of any manufacturing process must be proactive rather than reactive. Based on the knowledge that anomalous material is generally produced by interactions between different process variables, proactive reliability control reduces process variation and eliminates some of the failures that might occur in the future. Table 4-1 shows some important shifts of the reliability paradigm from traditional to new techniques.

For other issues surrounding CAD/CAM/CAPP systems, burn-in, life-cycle approach, and manufacturing yield and reliability, see Appendix A.

Manufacturing Yield and Reliability

In the past, most attempts to assure high IC reliability used product testing, life testing, or accelerated stress tests of the entire circuit. Because this approach to product testing is getting more expensive, more time consuming, and less able to identify the causes of parametric and functional failures of ICs, new approaches

TABLE 4-1 Reliability Paradigm for Nanoproducts

Item	Traditional Techniques	New Techniques
Approach	Reactive	Proactive
Test methodology	Output	Input
Failure mode	Separated	Integrated
Emphasis	Quality	Reliability

are needed. These new approaches make it possible to eliminate wear-out failures due to operational life.

There is a strong correlation between the number of field and life test failures and the manufacturing yield.[60,61] Kuper et al.[62] and van der Pol et al.[63] present models for the yield-reliability relation and experimental data to show the correlation. Thus, the root causes of reliability failures are the same as those of yield failures, and the manufacturing yield depends on the number of defects found during the manufacturing process, which in turn determines reliability.

The degree of manufacturing success is measured by yield, which is defined as the average ratio of devices on a wafer that pass the tests to the total number of devices on the wafer. Since the yield is a statistical parameter and implies a probability function, yield functions are multiplied to attain the total yield. The total wafer yield is a measure of good chips per wafer normalized by the number of chip sites per wafer. Generally, since yields can be measured in various ways, the overall yield is calculated as the product of elements of yield, such as the line yield, the wafer probe yield, the assembly yield, the final test yield, and the burn-in yield.

Parameters that affect yield and defects, and the number of defects produced during the manufacturing process can be effectively controlled by introducing testing at critical processing steps rather than throughout the assembly line.[64] This not only improves the reliability of the outgoing product but also significantly enhances the yield of the manufacturing process, thus increasing the quality of the overall system. Test points are effective only at critical processing steps, and their random distribution in the process was observed not to yield the desired results of high quality and minimal defects density. There is another way to control the yield. Since IC device yields are a function not only of chip area but also of circuit design and layout,[65,66] it is possible to control and manage the yield of ICs by determining the probabilities of failure and critical areas for different defect types.

COMMERCIALIZATION

Commercialization of a diverse set of products using a new technology is not the same as the commercialization of a new product. Many treatises have discussed the barriers to the commercialization of new products in great detail. For example, *The Innovator's Dilemma*[67] is an excellent text that analyzes the problems of bringing new technologies to bear on existing industries and the new commercial products that result.

The commercialization of a new technology that spans many different product categories, however, is an entirely different situation. Here, the issue is the use (presumably in multiple, diverse products and by multiple, diverse companies) of a particular set of manufacturing techniques that define the new technology. There are at least four prerequisites for effective commercialization of a new technology:

- Identification of existing products that will benefit from the new technology or of possibilities for creating new ones.
- Wide access to the technical details of the new technology.
- Enlightened corporate management.
- Sufficient reduction in product cost.

Identification of Products Manufactured in the New Technology

The identification of existing products that will benefit from the new technology or the creation of new products is an essential step in the acceptance and commercialization of a new manufacturing technology. Government can help industry meet this challenge, but only to the modest extent that the typically small government market can demand new products using the technology.

Wide Access to the Technical Details of the New Technology

The technical details of the new technology must be widely available so that a significant number of individuals are intimately acquainted with it or else the technology will never be developed. Initially, the only people familiar with the new technology are the research staff. A specific and deliberate effort must be made to transfer the expertise from research to product design. Expecting the design engineering community to discover and read the papers published by the research community is unrealistic. In the 1960s, the federal government, as the largest single customer for electronics, drove the migration in electronics, allowing companies to develop design engineering expertise at government expense (see Chapter 5). In today's climate, the government will purchase only a small fraction of the devices made possible by micro- and nanotechnology. Further, many high-technology companies are financially strained by small margins. It thus behooves the government to find a way to drive the education of the design engineering community in micro- and nanotechnology in order to guarantee that industry truly does benefit from the federal research investment.

Enlightened Corporate Management

Enlightened corporate management willing to embrace the new technology is a key accelerator. This is not, per se, a government issue. However, government programs can create the knowledge base and make available infrastructure that increases awareness of the possibilities of a new technology and lowers the barriers to early product development. An example is the MOSIS project (see Box 4-1) and DoD's role in the development of electronics. The DoD funded both research in new electronic technology and a foundry for making that new technology accessible to industry, government, and academia. MOSIS was a project funded to bring CMOS to as wide a community as possible by providing inexpensive, very-small-lot prototyping services. Few realize that Bill Joy used MOSIS[68] to fabricate a prototype processor as one of his efforts to help launch a then infant start-up, Sun Microsystems.

Sufficient Reduction in Product Cost

A significant investment of corporate resources usually must be made to climb the learning curve and drive down manufacturing costs. Although most of the responsibility for this critical step falls on industry, government can play a strong role by maintaining critical-mass investments in the refinement of the new manufacturing technology. This is precisely the history of CMOS. Once sufficient industrial interest in CMOS had been developed by the MOSIS program, individual companies developed their own versions of CMOS that had superior performance, better reliability, and lower cost than MOSIS CMOS.

Government Role in Providing Wide Access to New Technology

One excellent example of a successful government program that gives the design engineering community wide access to new microtechnology (in this case, MEMS) is the MEMS Exchange offered by the Center for National Research Initiatives and supported by DARPA.

The MEMS Exchange is a distributed foundry service for MEMS, funded in part by DARPA since 1998. Currently the MEMS Exchange offers nearly 800 processes from 22 independent fabrication sites around the country. Over 1,250 MEMS designers from nearly 200 business, academic, and government organizations access the MEMS Exchange for their fabrication needs. This program stands as an example of how government funding can create access to the technical details of a new technology.

The overall goals were to provide designers with a maximum of process and design freedom, with quick turnaround time, and with high-quality prototyping. To achieve these goals, a distributed MEMS processing environment composed

BOX 4-1
MOSIS

The Metal Oxide Semiconductor Implementation Service (MOSIS) was established in 1981 as a low-cost prototyping and small-volume production service for VLSI circuit development. MOSIS accepts designs from commercial firms, government agencies, and research and educational institutions around the world. It integrates user designs for a common process onto a single mask set, supplies the mask set to commercial foundries for fabrication, dices the finished wafers, and ships the individual dies to their corresponding requestors. This approach spreads the mask-making and fabrication costs among many users to make cost-effective fabrication of prototype chips possible. A single mask set can cost $50,000 and up, depending on the minimum line feature size and number of layers required. A 2.2-mm-square "tiny chip" fabricated using a 1.5-micrometer CMOS process costs only $1,080 for five copies. This low-cost fabrication service is ideal for students and researchers who need only a few copies of a given design or need to verify their designs before they enter mass production. MOSIS stimulates creativity in semiconductor circuit design and in related fields such as CMOS-compatible MEMS by offering a low-cost prototyping service. Figure 4-1-1 shows a sample detector die for a nano/picosatellite sun sensor fabricated as a "tiny chip." MOSIS currently offers CMOS fabrication with 1.5-, 0.5-, 0.35-, 0.25-, and 0.18-micron processes from various vendors, 0.5- and 0.25-micron SiGe BiCMOS processes, a 0.5-micron silicon-on-insulator process, and a 0.2-micron GaAs process. See <http://www.mosis.org> for more information.

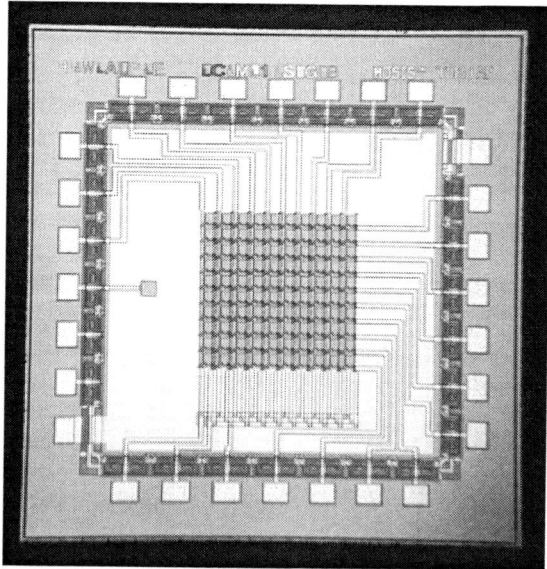

FIGURE 4-1-1 Two-dimensional active pixel sensor array. Photograph courtesy of the Aerospace Corporation.

of a number of separate fabrication sites organized in a fabrication network and made available to the community through a central organization, the MEMS Exchange, was organized. An important and unique element is that designers can have the process sequence for their devices conducted at multiple fabrication sites. This approach affords designers with an enormous range of choices in processing techniques and materials as well as the ability to fully customize the processing sequence.

Figure 4-11 shows the rapid growth in user accounts for the MEMS Exchange. Furthermore, over the past 3 years, distributed, multiple-site MEMS device fabrication has been shown not only to be feasible but also to produce advanced and complex prototype devices with distinct advantages over those made at a single foundry. In the last 2 years, nearly 400 process sequence runs were successfully performed in the MEMS Exchange network for designers around the country, each run being different and customized for a particular application. Most of the process sequences delivered by the MEMS Exchange were performed at two sites, with a surprising number of runs having had processing work performed at three or more sites. The designers accessing the MEMS

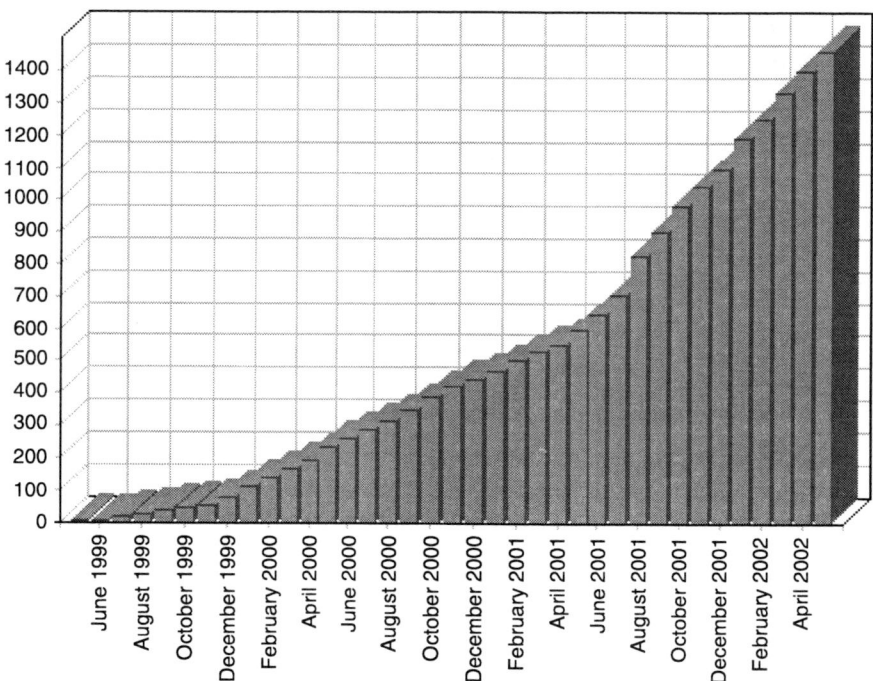

FIGURE 4-11 Cumulative user accounts for the MEMS exchange. Courtesy of the MEMS Exchange.

Exchange come from a wide range of organizations, including commercial (40 percent), academic (40 percent), and government laboratories (20 percent).

Fabrication sites enlisted in the MEMS Exchange include the University of California at Berkeley; Stanford University; Cornell University; the University of Michigan; Case Western Reserve University; Louisiana State University; the University of Illinois; Sony Semiconductor; Integrated Sensing Systems Corporation; Lance Goddard Associates; Microwave Bonding Corporation; Teledyne Electronics Corporation; Advanced MEMS Optical; Tactical Fabs, Inc.; Analog Devices; ASML; Zygo Teraoptics; Intelligent Micropatterning; Axsun Technologies; Aspen Technologies; Fiberlead, Inc.; and American Precision Dicing. Through these 22 fabrication sites that belong to the MEMS exchange, an extremely wide range of MEMS fabrication resources are available to the community, and new process capabilities are being added each week.

Effect of Manufacturing Complexity on Commercialization

Government-sponsored technology foundry programs (such as MOSIS, the Microelectronics Center of North Carolina, and the MEMS Exchange) have been successful in providing some level of technology access to the broader community, although demand for the services usually far outstrips capacity. However successful these programs may be in providing wide access to a technology, the cost barriers to the adoption of a new technology can only be solved by industry. Thus, government must make sure to involve industry in the new technology as soon as possible.

Manufacturing complexity strongly affects manufacturing cost. Thus, for technologies such as MEMS, cost depends on device complexity—in effect, how much integration between IC electronics and micromechanics has been accomplished. The three levels of manufacturing complexity of MEMS are these:

- all micromechanics and no IC electronics
- multichip modules or other assembled hybrids
- integrated micromechanics and IC electronics

Commercialization has already occurred for MEMS on all three levels of complexity.

The third level is the most complicated and requires the most fabrication sophistication from industry. This level, in turn, can be subdivided into three specialties:

- preprocessed micromechanics, postprocessed IC electronics
- integrated micromechanics and IC electronics
- postprocessed micromechanics, preprocessed IC electronics

Case Study: Texas Instruments and the Digital Mirror Device

Commercialization does not necessarily happen earliest for the simplest forms of manufactured MEMS (either those containing no IC electronics or hybrids of micromechanics and IC electronics). As an example of how one of the more difficult manufacturing processes (integrated micromechanics and IC electronics) was used for MEMS commercialization, the story of the Texas Instruments (TI) digital mirror device [DMD™] is related below. It shows that even after DMD feasibility had been demonstrated in the late 1980s, the commercialization effort required more than two decades to achieve profitability. This effort took place before the advent of government programs to lower the barriers to commercialization of MEMS. The cost of the DMD and the associated projector, called the digital light processor (DLP), is reported to have been nearly a billion dollars. Other commercialization efforts, most notably the MEMS airbag crash detection accelerometer, manufactured by Analog Devices, Inc. (ADI) had similar difficulties, but commercialization was accomplished at much lower cost and in a shorter time. This was primarily due to the relatively simpler structures ADI had to fabricate but also to the influx of engineers trained in MEMS in government programs.

TI was a pioneering company, well in advance of the technology curve, and was forced to solve many newly discovered MEMS problems entirely on its own. For micro- and nanotechnology to be commercialized quickly, it will be necessary for government not only to continue pursuing research in these technologies but also to guarantee wide access to them through programs specially funded for this purpose. The government must ask itself, first, if it really can expect most companies to be persistent in the face of product development adversity and, second, if it can wait so long for the new products to arrive.

TI's DMD, shown in Figures 4-12 and 4-13, consists of an array of a half million or more mirrors that can electrically switch light to a working area. Figure 4-12 shows a cut-away of the DMD structural model. Figure 4-13 shows a photomicrograph of the DMD. The levels labeled on the photomicrograph are identified in Table 4-2 illustrating the complexity of the DMD, with seven functions squeezed into three levels.

The original DMD started out as an array of polymer film mirrors originated by a National Security Agency research group whose mission was optical correlation processing. Under DARPA support the mirror array technology was transferred to TI for development. Each mirror had to have a charge placed under the membrane to cause deflection and thus the enable the parallel correlation operations. This called for a large array of semiconductor circuits to enable the deflection. Feasibility was demonstrated, but manufacturing problems—for example, those from dust particles under the membrane causing defective mirror pixels—quickly emerged.

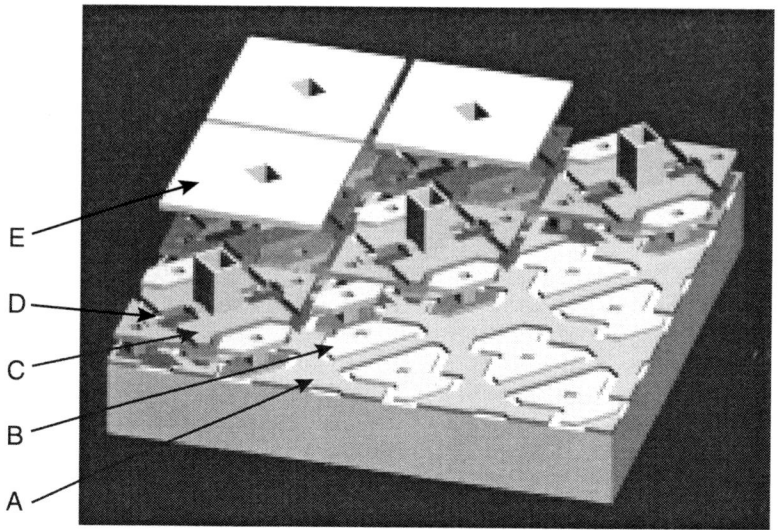

FIGURE 4-12 Cut-away of the digital mirror device structural model. Reprinted with permission of Michael Mignardi, Texas Instruments.

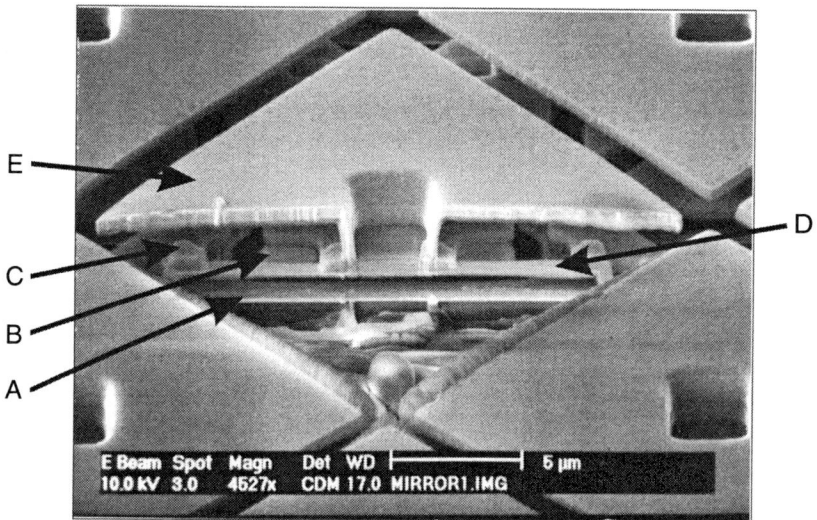

FIGURE 4-13 Photomicrograph of the digital mirror device. Reprinted with permission of Michael Mignardi, Texas Instruments.

TABLE 4-2 High Complexity of the Digital Mirror Device

Label	DMD Level	Function
A	Electrode	Bias Electrode
B	Electrode	Address Electrode
C	Hinge/Beam	Beam and Hinge Posts
D	Hinge/Beam	Hinges
E	Mirror	Mirror and Mirror Support Post

A monolithic mirror approach using an array of aluminum alloy mirrors was developed to overcome the particle problems of the polymer mirrors. The aluminum mirrors worked much better in regard to dust but rapidly failed in operation because of the stress inherent in the mirror layer after plasma etching. Also, the TI printer division interested in the printer application fell on hard times, so the continuation of the program was always in doubt.

New DARPA money was supplied, administered by the Air Force. DARPA was then interested in large projection displays. During the 1990s development problems continued. (For instance, in operation, the mirrors would stick permanently to the underlying films; the mirrors did not optimally fill the optical aperture, leading to a checkerboard display image; there was high optical insertion loss from a mismatch in the pixel size and the size of the arc lamp; a very high light flux on the DMD was required; arc lamp longevity was insufficient; and the cost of engineering of a small sophisticated light handling system, including an optically flat window on the package top, was high.) By the late 1980s the DMD was being touted as a MEMS success story, but these manufacturability issues continued to threaten the viability of commercialization.

During the 1990s, TI, using its own funds, gradually worked out solutions to the DMD manufacturability problems. Mirror reliability was solved with a hidden mirror hinge made out of an amorphous material. Since there were no grain boundaries in the hinge, there were no sources of the cracks that had led to high failure rates in the metal hinges. Stiction was fixed by fabricating springs and a fluorocarbon coating on the back side of the mirrors. The springs produced a restoring force when the mirror touched the base layer. The hidden-hinge design allowed for a more optically efficient pixel, filling the aperture. Obtaining a robust, very bright light source turned out to be an enormous problem, and the light source is still a major cost component in the DMD system.

One of the veteran engineers on the project at Texas Instruments, Michael Mignardi, recently gave an IEEE seminar detailing the scale of TI's effort to successfully produce the DMD.[69] Even after feasibility had been demonstrated in the late 1980s, it took another 10 years to reach profitability, with the effort being spread roughly equally among device improvement, packaging, and testing. Low outgasing adhesives were used to prevent the buildup of a sticky layer inside the package, which would result in stiction problems. A zeolite getter was included in the package to mitigate water-induced stiction in the completed package. These and other innovations led to TI turning its first profit on projectors in 2000.

Specifications for a 1.2-million-pixel projector device include mirrors that tilt 10 to 12 degrees to deflect light onto a screen. Only five stuck-on pixels are permitted for a shipped DMD. Clearly, achieving this level of yield and reliability required an enormous effort in developing test equipment and lifetime testing for the semiconductor layer, the mirror level, and the combined levels in action.

TI and its technological team deserve a special place in the MEMS Hall of Fame for their perseverance in bringing the projector product to market. The Air Force also deserves credit for early support. Most importantly the development of the successful DMD back-end processing indicates that such processing can be solved by a determined team, even for very complex devices and systems. But it also illustrates the very real difficulties in turning a laboratory demonstration into a reliable product—difficulties that will undoubtedly confront all emerging micro- and nanotechnology applications in as yet unanticipated ways.

New technologies are generally thought to become profitable after 10 years. It is generally believed that so far few, if any, MEMS projects have become profitable long-term. All of which is to say that building micro- and nanotechnology materials and systems is very difficult. Continuing to approach these technologies in the cottage industry mode that is currently fashionable will not produce DMD-type successes. There is a need for farsighted management and technical staffs—and for deep pockets. Further, the multidisciplinary research capacity that TI had available does not exist in many of today's companies in the face of the increased pressure for short-term profitability, and it certainly is not possible in the small-company entrepreneurial environment that is attempting to bring nanotechnology to fruition.

FINDINGS AND RECOMMENDATIONS

Finding 4-1. *Lithography and pattern transfer and self-assembly are key enablers for evolving micro- and nanotechnologies.*

Recommendation 4-1. *The AFRL R&D program will require access to micro- and nanolithography and pattern transfer tools. This should be accomplished using available national facilities or otherwise providing the function internally. Research into new nanolithography and patterning technologies, complementary to the industry push for high-throughput tools, would be a worthwhile investment.* The Air Force should not compete with industry efforts, particularly in silicon technology, but should concentrate on developing processes for structures and materials that are outside traditional silicon processing—for example, deep etching for MEMS and integration of new materials with silicon.

Finding T7. *Integration of micro- and nanoscale processes and of different material systems will be broadly important for materials, devices, and packaging. Self-assembly and directed assembly of dissimilar elements will be necessary to maximize the functionality of many micro- and nanoscale structures, devices, and systems. Achievement of high yields and long-term reliability, comparable to those of the current integrated circuit industry, will be a major challenge.*

Recommendation T7. *The Air Force should monitor progress in self- and directed-assembly research and selectively invest its R&D resources. It will be critical for the Air Force to participate in developing manufacturing processes that result in reliable systems in technology areas where the military is the dominant customer—for example, in sensors and propulsion systems.* Developments in many of these areas will be driven by the commercial sector. The Air Force must stay aware of advances and apply them to its unique needs. As an example, in sensor applications a wide range of otherwise incompatible materials and fabrication processes is likely to be necessary.

Finding 4-2. *So much is already known about progress in silicon, with its already highly developed and constantly improving manufacturing processes, that it is unlikely a sui generis technology will spring up sufficiently developed and robust that it will immediately supplant not only the transistor but also all the rest of the integrated circuit.* Integrated circuit technology has become extremely sophisticated, and the industry is devoting extensive resources to extending this sophistication in its drive to validate Moore's law for future IC generations. In contrast, nanotechnology is at a much earlier development stage, concentrating on the behavior of individual devices and circuit components (switches, wires, etc.). It is most likely that these new technologies will first find use as complements to silicon, not as immediate replacements for integrated circuits. Over the longer term, it is not possible to predict the relative roles of integrated circuits and new and evolving nanotechnologies.

Recommendation 4-2. *The Air Force should emphasize those areas of micro- and nanotechnology for information processing that are potentially integrable with silicon technology and that address Air Force-specific, noncommercial military applications.*

Finding 4-3. *The path from laboratory demonstration to the manufacture of reliable devices and systems is long and arduous, requiring extensive resources and prodigious technology development.* This will undoubtedly be as true of today's emerging technologies as it has been throughout the history of technology. It is worthwhile to consider this lesson when listening to the siren songs appearing daily, particularly in the popular and business press, on the future benefits of nanotechnology. There is undoubtedly an exciting future, and just as undoubtedly, we will find many surprises, both positive and negative, along the way.

Recommendation 4-3. *Air Force research efforts should be directed not only to the science of micro- and nanotechnology, but also to the development of devices and systems and wide access to the manufacturing technology required to produce them.*

REFERENCES

1. Nagel, D.J. 2002. Technologies for micrometer and nanometer pattern and material transfer. Pp. 557–701 in Direct-Write Technologies for Rapid Prototyping Applications: Sensors, Electronics, and Integrated Power Sources. A. Pique and D.B. Chrisey, eds. New York, N.Y.: Academic Press.
2. Bjorkholm, J.E. 1998. EUV lithography–the successor to optical lithography? Available online at <http://www.intel.com/technology/itj/q31998/articles/art_4.htm> [July 8, 2002].
3. Smith, H.I. 2001. Japan could dominate industry with x-ray lithography. Available online at <http://www.e-insite.net/semiconductor/index.asp?layout=article&articleId=CA61809&stt=001> [July 8, 2002].
4. Institut für Mikrotechnik Mainz. 2001. LIGA Technology. Available online at <http://www.imm-mainz.de/> [July 8, 2002].
5. Chou, S.Y., L. Zhuang, and L.J. Guo. 1999. Lithographically induced self-construction of polymer microstructures for resistless patterning. Applied Physics Letters 75(7): 1004–1006.
6. Chou, S.Y., and L. Zhuang. 1999. Lithographically induced self-assembly of periodic polymer micropillar arrays. Journal of Vacuum Science Technology B 17(6): 3197–3202.
7. Becker, H., and U. Heim. 1999. Silicon as tool material for polymer hot embossing. Pp. 228–231 in Proceedings of MEMS '99: The 12th IEEE International Conference on Micro Electro Mechanical Systems. New York, N.Y.: IEEE.
8. Glinsner, T. 2001. Nanoimprinting can solve pattern-generation problems. R&D Magazine 43(7): 33.
9. Chou, S.Y., P.R. Krause, and P.J. Renstrom. 1995. Imprint of sub-25 nm vias and trenches in polymers. Applied Physics Letters 67(21): 3114–3116.
10. Chou, S.Y. 2001. Nanoimprint lithography. Available online at <http://www.ee.princeton.edu/~chouweb/newproject/page3.html> [July 8, 2002].
11. Johnson, S.C. 1999. Selective Compliant Orientation Stages for Imprint Lithography. Available online at <http://sfil.org/research/papers/sjthesis.pdf> [July 8, 2002].
12. Xia, Y., and G.M. Whitesides. 1998. Soft lithography. Annual Review of Materials Science 28: 153–184.
13. Senturia, S.D. 2001. Microsystem Design. Boston, Mass.: Kluwer Academic Publishers.
14. Kauffman, S.A. 1995. At Home in the Universe: The Search for Laws of Self-Organization and Complexity. New York, N.Y.: Oxford University Press.
15. This series, for example, provides a good overview of the topic: Lehn, J.M., J.L. Atwood, J.E.D. Davis, D.D. MacNicol, and F. Vögtle. 1996. Comprehensive Supramolecular Chemistry, Vols. 1–11. New York, N.Y.: Pergamon.

16. Lehn, J.M. 1995. Supramolecular Chemistry: Concepts and Perspectives. New York, N.Y.: VCH Press.
17. Balzani, V., and L. DeCola. 1992. Supramolecular Chemistry. Dordrecht, The Netherlands: Kluwer Academic Press.
18. Desiraju., G.R. 1996. The Crystal as Supramolecular Entity: Perspectives in Supramolecular Chemistry, Volume 2. New York, N.Y.: John Wiley & Sons.
19. Desiraju, G.R. 1989. Crystal Engineering: The Design of Organic Solids. New York, N.Y.: Elsevier.
20. Manne, S., and G.G. Warr. 1999. Supramolecular Structure in Confined Geometries. Washington, D.C.: American Chemical Society.
21. Jones, M.N., and D. Chapman. 1995. Micelles, Monolayers and Biomembranes. New York, N.Y.: Wiley-Liss.
22. Wilcoxon, J.P., J.E. Martin, and D.W. Schaefer. 1989. Aggregation in colloidal gold. Physical Review A 39(5): 2675–2688.
23. Kumar, A., N.L. Abbott, E. Kim, H.A. Biebuyck, and G.M. Whitesides. 1995. Patterned self-assembled monolayers and mesoscale phenomena. Accounts of Chemical Research 28(5): 219–226.
24. De Gennes, P.G., and J. Prost. 1993. The Physics of Liquid Crystals, 2nd Ed. New York, N.Y.: Oxford University Press.
25. Rouvray, D. 2000. Molecular self-assembly. Chemistry in Britain 36(7): 26–29.
26. Huc, I., and J.M. Lehn. 1997. Virtual combinatorial libraries: Dynamic generation of molecular and supramolecular diversity by self-assembly. Proceedings of the National Academy of Sciences 94(6): 2106–2110.
27. Frechet, J.M.J. 2002. Dendrimers and supramolecular chemistry. Proceedings of the National Academy of Sciences 99(8): 4782–4787
28. Whitesides, G.M., and J.C. Love. 2001. The art of building small—Researchers are discovering cheap, efficient ways to make structures only a few billionths of a meter across. Scientific American 285(3): 38–47.
29. Van Hest, J.C.M., and D.A. Tirrell. 2001. Protein-based materials: Toward a new level of structural control. Chemical Communications (19): 1897–1904.
30. Yu, S.J.M., C.M. Soto, and D.A. Tirrell. 2000. Nanometer-scale smectic ordering of genetically engineered rodlike polymers: Synthesis and characterization of monodisperse derivatives of poly (g-benzyl a, L-glutamate). Journal of the American Chemical Society 122(28): 6552–6559, and references therein.
31. Limberis, L., and R.J. Stewart. 2000. Toward kinesin-powered microdevices. Nanotechnology 11(2): 47–51.
32. Srinivasarao, M. 1999. Nano-optics in the biological world: Beetles, butterflies, birds, and moths. Chemical Reviews 99(7): 1935–1961.
33. Miller, J.S., and A.J. Epstein. 2000. Molecule-based magnets—An overview. MRS Bulletin 25(11): 21–28.
34. Ovcharenko, V.I., and R.Z. Sagdeev. 1999. Molecular ferromagnets. Russian Chemical Reviews 68(5): 345–363.
35. Alivisatos, A.P. 2001. Less is more in medicine—Sophisticated forms of nanotechnology will find some of their first real-world applications in biomedical research, disease diagnosis, and, possibly, therapy. Scientific American 285(3): 67–73.
36. Hooks, D.E., T. Fritz, and M.D. Ward. 2001. Epitaxy and molecular organization on solid substrates. Advanced Materials 13(4): 227–241, and references therein.
37. Choquette, K.D., K.L. Lear, R.P. Schneider, K.M. Geib, J.J. Figiel, and R. Hull. 1995. Fabrication and performance of selectively oxidized vertical-cavity lasers. IEEE Photonics Technology Letters 7(11): 1237–1239.

38. Hatzor, A., and P.S. Weiss. 2001. Molecular rulers for scaling down nanostructures. Science 291(5506): 1019–1020.
39. For a contemporaneous snapshot of this field, see, for example, Fan, J.C.C., and J.M. Poate. 1986. Heteroepitaxy on Silicon. Materials Research Society Symposia Proceedings 67. Pittsburgh, Pa.: Materials Research Society.
40. Kapon, E., D.M. Hwang, and R. Bhat. 1989. Stimulated-emission in semiconductor quantum wire heterostructures. Physical Review Letters 63(4): 430–433.
41. Zhou, Y.X., S. Luong, C.P. Hains, and J. Cheng. 1998. Oxide-confined monolithic, multiple-wavelength vertical-cavity surface-emitting laser arrays with a 40-nm wavelength span. IEEE Photonics Technology Letters 10(11): 1527–1529.
42. For a recent survey, see the special section on the growth of heterostructure materials on nanoscale substrates, S.D. Hersee, D.A. Zubia and S.R.J. Brueck, eds. 2002. IEEE Journal of Quantum Electronics 38(8).
43. Cui, Y. and C.M. Lieber. 2001. Functional nanoscale electronic devices assembled using silicon nanowire building blocks. Science: 291(5505): 851–853.
44. Duan, X.F., Y. Huang, Y. Cui, J.F. Wang, and C.M. Lieber. 2001. Indium phosphide nanowires as building blocks for nanoscale electronic and optoelectronic devices. Nature 409(6816): 66–69.
45. Gudiksen, M.S., L.J. Lauhon, J. Wang, D.C. Smith, and C.M. Lieber. 2002. Growth of nanowire superlattice structures for photonics and electronics. Nature 415(6872): 617–620.
46. International Technology Roadmap for Semiconductors. 2001. Available online at <http://public.itrs.net/> [July 3, 2002].
47. Moravec, H. 1998. When will computer hardware match the human brain? Available online at <http://www.transhumanist.com/volume1/moravec.htm> [July 8, 2002].
48. Talghader, J.J., J.K. Tu, and J.S. Smith. 1995. Integration of fluidically self-assembled optoelectronic devices using a silicon-based process. IEEE Photonics Technology Letters 7(11): 1321–1323.
49. Esener, S.C., D. Hartmann, M.J. Heller, and J.M. Cable. 1998. DNA-assisted microassembly: A heterogeneous integration technology for optoelectronics. Paper # CR70-07 in Heterogeneous Integration: Systems on a Chip, Proceedings of SPIE, Volume CR70. A. Husain and M. Fallahi, eds. Bellingham, Wash.: The International Society for Optical Engineering.
50. Hartmann, D.M., M. Heller, S.C. Esener, D. Schwartz, and G. Tu. 2002. Selective DNA attachment of micro- and nanoscale particles to substrates. Journal of Materials Research 17(2): 473–478.
51. Ozkan, M., C.S. Ozkan, O. Kibar, and S.C. Esener. 2000. Massively parallel low-cost pick-and-place of optoelectronic devices by electrochemical fluidic processing. Optics Letters 25(17): 1285–1287.
52. Ozkan, M., C.S. Ozkan, O. Kibar, M.M. Wang, S. Bhatia, and S.C. Esener. 2001. Heterogeneous integration through electrokinetic movement. IEEE Engineering in Medicine and Biology Magazine 20(6): 144–151.
53. Hartmann, D.M., O. Kibar, and S.C. Esener. 2001. Optimization and theoretical modeling of polymer microlens arrays fabricated using the hydrophobic effect. Applied Optics 40(16): 2736–2746.
54. Hartmann, D.M., D.J. Reiley, and S.C. Esener. 2001. Microlenses self-aligned to optical fibers fabricated using the hydrophobic effect. IEEE Photonics Technology Letters 13(10): 1088–1090.
55. Harper, C.A. 2000. Electronic Packaging and Interconnection Handbook, Third Edition. New York, N.Y.: McGraw-Hill.
56. Gilleo, K. 2002. Area Array Packaging Handbook. New York, N.Y.: McGraw-Hill.
57. Kuo, W., W.T.K. Chien, and T. Kim. 1998. Reliability, Yield, and Stress Burn-In: A Unified Approach for Microelectronics Systems Manufacturing and Software Development. Boston, Mass.: Kluwer Academic Publishers.

58. Kim, T., W. Kuo, and W.T.K. Chien. 2000. Burn-in effect on yield. IEEE Transactions on Electronics Packaging Manufacturing 23(4): 293–299.
59. Shideler, J.A, T. Turner, J. Reedholm, and C. Messick. 1995. A systematic approach to wafer level reliability. Solid State Technology 38(3): 47–54.
60. Kim, T., W. Kuo, and W.T.K. Chien. 2000. Burn-in effect on yield. IEEE Transactions on Electronics Packaging Manufacturing 23(4): 293–299.
61. Kim, T. and W. Kuo. 1999. Modeling manufacturing yield and reliability. IEEE Transactions on Semiconductor Manufacturing 12(4): 485–492.
62. Kuper, F., J. van der Pol, E. Ooms, T. Johnson, R Wijburg, W. Koster, and D. Johnston. 1996. Relation between yield and reliability of integrated circuits: Experimental results and application to continuous early failure rate reduction programs. Pp. 17–21 in Proceedings of the 1996 34th Annual IEEE International Reliability Physics Symposium. New York, N.Y.: Institute of Electrical and Electronics Engineers.
63. Van der Pol, J., F. Kuper, and E. Ooms. 1993. Relation between yield and reliability of integrated circuits and application to failure rate assessment and reduction in the one digit fit and ppm reliability era. Microelectronics and Reliability 36(11/12): 1603–1610.
64. Shideler, J.A, T. Turner, J. Reedholm, and C. Messick. 1995. A systematic approach to wafer level reliability. Solid State Technology 38(3): 47–54.
65. Kuo, W., and T. Kim. 1999. An overview of manufacturing yield and reliability modeling for semiconductor products. Proceedings of the IEEE 87(8): 1329–1344.
66. Kim, T., and W. Kuo. 1999. Modeling manufacturing yield and reliability. IEEE Transactions on Semiconductor Manufacturing 12(4): 485–492.
67. Christensen, C.M. 2000. The Innovator's Dilemma: When New Technologies Cause Great Firms to Fail. New York, N.Y.: HarperBusiness.
68. Personal Communication between Keith Uncapher, Dean Emeritus, University of Southern California, and Al Pisano, Committee on Implications of Emerging Micro and Nano Technologies, March 2001.
69. Mignardi, M. 2001. The Digital Micromirror Device: Its Fabrication and Use. Presentation by Michael Mignardi Texas Instruments, to the IEEE weekly seminar, Southern Methodist University, Dallas, Texas, November 27.

5

Air Force Micro- and Nanotechnology Programs and Opportunities

IMPACTS OF MICRO- AND NANOTECHNOLOGIES ON AIR FORCE MISSIONS

The committee found four major themes that describe the impacts that micro- and nanotechnologies will have as they provide revolutionary advances for Air Force missions:

- increased information capabilities
- miniaturization of systems
- new materials resulting from new science at these scales
- increased functionality and autonomy

The dramatic increase in information capabilities of the past several decades—processing, storing, communicating, and displaying—shows no signs of imminent slowing. The integrated circuit industry is predicting ~128× improvements in transistor density, based on current state of the technology, over the next 15 years or so. Many other technologies are being explored for use once CMOS silicon scaling has reached saturation (see "Information Technology," the first section of Chapter 3, for additional discussion). This will allow the Air Force to pack more and more intelligence into smaller and smaller, lighter and lighter, less-power-consuming (per function) packages that have increasing local intelligence and increasing autonomy.

For space systems the ability to miniaturize will allow smaller and lighter vehicles, increased quality control, and new options for launch. Both space systems and avionics will benefit from the associated trend to modularization and standardization with improved reliability and better control of development costs. The adoption of the batch manufacturing and built-in-place fabrication approaches brought about by microtechnologies will increase reliability and reduce cost and will be especially enabling. The continued rapid advance of information technology will make possible new systems approaches. Miniaturization is expected to bring a world of ubiquitous sensing and computing to Air Force systems (lower power, smaller size, more function with embedded systems, lower cost, and increased capability), and the concomitant development of software for platforms, sensors, and C3I will become increasingly important. The challenge will be to discover ways to optimize the benefit from these rapidly evolving areas of technology.

Properties qualitatively change as things get small. There is great opportunity for discovering the new science and material properties that can be achieved at the micro- and nanoscale and for applying these capabilities to Air Force systems. The new properties being explored at the small scale, particularly the nanoscale, will lead to new capabilities for sensing, information processing and storage, propulsion, and high-performance materials. Stronger, more durable, and lighter-weight structures and increased efficiency in the use of energy are of particular interest. A particularly important challenge is the increased use of smart and adaptive materials—for example, to improve boundary layer control on aircraft or allow reconfigurable surveillance systems that learn from and adapt to their environment—to enable new systems approaches. Over the long term, these advances are likely to be accelerated as the study of biologically inspired systems leads to new ways to control, tailor, and adapt the properties of materials to their environment and to more efficiently store and utilize energy and information in small-scale systems.

The advances in information density, miniaturization, and materials functionality will enable an advanced degree of autonomous systems operation and a paradigm shift from reliance on a few large systems to many small things that work together. Over time, the increased intelligence of unmanned systems with integrated sensor suites and information processing (situational awareness) capabilities, and offensive/defensive reaction capabilities will probably change the character of warfare, allowing a degree of autonomy unthinkable today. The opportunities made possible by understanding and exploiting the emergent behavior of large numbers of entities working together as a system is a long-term challenge that will lead to new system architectures and revolutionary capabilities in analogy to biological systems.

CURRENT INVESTMENTS BY THE AIR FORCE IN MICRO- AND NANOTECHNOLOGIES

The Air Force, like the other DoD services and agencies, has historically invested in micro- and nanotechnologies. This investment supports both development-oriented efforts (6.2 and 6.3) at AFRL laboratories and 6.1 basic research through the Air Force Office of Scientific Research (AFOSR), a directorate within the Air Force Research Laboratory (AFRL).

AFRL Research Portfolio in Micro- and Nanotechnologies

In microtechnology, AFRL supports 40 projects for a total of $29 million. The AFRL nanotechnology program includes about 120 tasks, which receive over $25 million in support. This includes about $10 million for the support of nanomaterials research, about $12 million for nanodevice research, about $2 million for research in nanobiology and information technology, and about $3 million for research in nanoenergetics. This research is expected to have significant impact in the areas indicated in Box 5-1.

The relationship of the AFRL nanotechnology program to long-term Air Force requirements is illustrated in Table 5-1. Work is under way at the level of the AFRL Chief Technologist to collect the multiple existing threads in micro- and nanotechnology within AFRL and to plan a coordinated approach.

Funding for the main Air Force areas of interest by funding level is shown in Table 5-2 and in Figure 5-1.

AFOSR Basic Research Programs in Nanotechnology

Much of the basic research (6.1) is conducted in U.S. academic institutions; the distribution of AFOSR funding is about 70 percent external and 30 percent in-house within AFRL. These investments have been provided from the Air Force

---

BOX 5-1
Expected Impacts of Research Supported by the Air Force Nanotechnology Program

Sensors	Electronic devices	Energetic Materials	Enhanced Mechanical Structures
• Infrared target recognition • Airborne and space-based long-range detection • Multispectral awareness	• High-speed information processing • Orders of magnitude increase in computing power • Counterradiation effects	• Propellants with higher specific impulse and controlled burn rates • Smaller munitions • Safer propellants • Enhanced power generation • Advanced fuels, lubricants, and additives	• Lightweight structures • High performance • High-temperature materials and structures • Self-healing structures • Smart skins • Reduced cost of launch

TABLE 5-1 Challenges and Impact Areas

Long-Term Challenges	Areas Where Micro-and Nanotechnologies Will Have an Impact
Finding and tracking	*Nanosensors.* Integrated nanoelectronics and nanophotonics; enhanced infrared recognition; high-speed image processing
Command and control	*Nanodevices.* Nanoprocessors with orders of magnitude increase in computing power, information storage and processing abilities; radically improved decision making. *Quantum computing.* Eliminate multiple design iterations and prototype testing, extremely fast image reconstruction.
Controlled effects	*Nanoscale energetic materials.* Improved energy release rate; accelerated burn; smaller munitions; safer propellants. *Nanoelectronics.* Counter radiation effects.
Sanctuary	*Nanosensors.* Airborne and space-based long-range detection. *Coatings.* Revolutionary dynamic stealth.
Effective aerospace persistence	*Nanoparticles and nanostructured materials.* Advanced fuels, lubricants, and additives; power generation, storage, and delivery; long-life high-temperature components; self-healing structures; smart skins. *Nanoelectronics.* Nanosatellite clusters.
Rapid aerospace response	*Nanocomposites and nanostructures.* Lightweight structures; reduced cost to launch; high-performance and high-temperature materials and structures; high-efficiency propellants.

SOURCE: Adapted from Schöne, H. 2001. The Role of Nano-Technology and Microsystems for Space. Briefing by Harald Schöne, Air Force Research Laboratory Space Vehicles Directorate, to the Committee on Implications of Emerging Micro and Nano Technologies, Doubletree Hotel, Albuquerque, N.M., November 8.

core program, primarily through the Air Force Office of Scientific Research (AFOSR), and in collaboration with other DoD organizations such as DARPA and the Office of the Director of Defense Research and Engineering (DDR&E). The Air Force serves as an agent for DARPA, DDR&E, and other DoD agencies and manages a significant customer budget. These funds are leveraged against core Air Force funds to produce a strong and robust research program.

The Air Force also collaborates with other DoD and non-DoD federal agencies in defining and supporting a broad research program in micro- and nanotechnologies. It was a key participant in the Joint Service Electronics Program (JSEP), which supported a variety of electronics-oriented technologies. JSEP was instrumental in supporting research in compound semiconductor materials and devices and helped develop the technology for fabricating and evaluating sub-

TABLE 5-2 Air Force Nanotechnology Research

Area	Funding ($ millions)
Nanodevices	12
Nanomaterials	10
Nanoenergetics	3
Nanobiology/IT	2

SOURCE: Adapted from data provided by Gernot Pomrenke, Air Force Research Laboratory.

micrometer devices. JSEP research helped provide the tools and techniques that now make it possible to work on the nanoscale.

The JSEP program was phased out in 1997 primarily due to the growth of the University Research Initiative (URI). The Air Force has been a major participant in the URI since its establishment in 1986. URI was established to fund large, multidisciplinary group research programs in U.S. academic institutions. It is funded and managed by the Basic Research Office in DDR&E, but executed through the AFOSR, the Office of Naval Research (ONR), the Army Research Office (ARO), and DARPA. The URI now consists of a variety of programs, including the Multidisciplinary University Research Initiative (MURI) and the Defense University Research Instrumentation Program (DURIP). These programs account for over half the funds allocated to the URI and provide in turn a significant amount of university research funding to the Services. The MURI programs

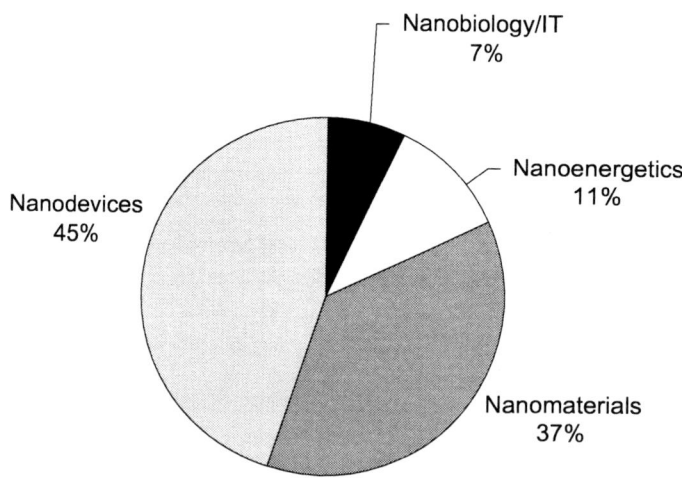

FIGURE 5-1 Air Force nanotechnology research. SOURCE: Based on data provided by Gernot Pomrenke, Air Force Research Laboratory.

are funded at about $1 million per year and there are currently about 140 programs in existence. Many of these programs are directed at nanotechnology and are providing a significant amount of funding for nanotechnology research. The DURIP has a budget of about $45 million. It is the only program funded by DoD that explicitly provides resources to U.S. academic institutions for acquiring large, expensive equipment items that cannot be supported by traditional grant mechanisms. Examples include molecular beam epitaxial semiconductor growth machines, organo-metallic chemical vapor deposition reactors, scanning tunneling electron microscopes, atomic force microscopes, and other equipment that enables research on nanoscale materials and devices. This support is leveraged by the Services with their core programs to support service-specific research, as well as work of general and broad interest.

The Air Force has been a participant in the establishment of the National Nanotechnology Initiative (NNI). The NNI is a multiagency program to provide government support for nanoscience and nanotechnology research as discussed in Chapter 2. The program was planned by a consortium of representatives from six U.S. government agencies. The original agencies were the National Science Foundation; the Departments of Defense, Commerce, and Energy; the National Institutes of Health; and the National Aeronautics and Space Administration. The program was proposed by the President in his 2001 budget request and funded by Congress. Each participating agency reviewed then-current research programs and determined that funding for nanotechnology research had been about $270 million in FY 2000. The NNI brought about an increase of 73 percent in the FY 2001 budget, for a total of $466 million. Additional increases that brought the total investment to over $600 million were included in the FY 2002 budget. Each agency surveyed its scientific and program offices to determine opportunities and areas of interest for nanotechnology research and to avoid overlapping efforts. The NNI representatives coordinated the survey results and turned them into a comprehensive program. Initially, DoD interest in nanotechnology focused on three major areas, as outlined in Box 5-2.

The Air Force has further reviewed its efforts in nanotechnology in order to refine the focus on areas more specific to its interests (Box 5-3).

In FY 2001, the MURI program conducted a nanotechnology-focused competition, Defense University Research in Nanotechnology (DURINT), with a total annual funding level of $8.75 million. The AFOSR-managed DURINT programs included two programs in nanodevices, three in nanomaterials, and one each in nanoenergetics and nanobiology/information technology (Table 5-3). In addition, the Air Force supported two nanotechnology-related MURIs in FY 2001 (Table 5-4). Overall, 5 of 48 MURI awards were in nanotechnology-related areas.

The FY 2001 DURINT program included a separate equipment competition similar to the DURIP. The AFOSR grants related to nanotechnology are shown in Table 5-5.

BOX 5-2
Initial DoD Focus in Nanotechnology

Information Acquisition, Processing, Storage, and Display

- Higher-speed electronics
- Higher-density electronics
- Lower-power electronics
- Increased complexity on a chip
- Optoelectronic capability

Materials Performance and Affordability

- Extended life and maintenance
- "Smart" materials
- High-performance materials

Nanobioengineering

- Chemical and biological agents
- Detection/destruction
- Casualty care (miniature devices to sense/actuate)
- Personnel health monitors/stimulators

BOX 5-3
Air Force Nanotechnology Program

Nanoengineered Materials

- Carbon nanotubes and composites
- High-temperature and high-strength materials and coatings
- Nanocomposites, organic and inorganic
- Multilayer laminates
- Self-healing polymers
- Self-assembly and hybrid fabrication
- Designer substrates for electronics
- Nanocontrolled dielectrics

Nanostructured Devices

- Ultrafast, ultradense electronic devices and processors
- Nanoscale sensors and emitters
- Molecular electronics and architecture
- Nanophotonics and optical nanoprobes
- Nonlinear and adaptive nanoscale optics
- Quantum computing devices and circuits
- Chemical and biological quantum sensors

Nano/Bio/Info Interface

- Nanobiocatalysis
- Chemical and biological decontamination
- Nanosystems architecture
- DNA information processing
- Biocomputational models
- Neural network processors
- Distributed systems

Nanoenergetics

- Nanoscopic fuel additives
- Nanoscale energetic materials
- High-energy-density materials
- Nanoscale photovoltaics
- Nanofuels, nanocomposites
- Nanofluidics and plasma aerodynamics
- Unimolecular micelles
- Laser sources

TABLE 5-3 AFOSR-Managed DURINT Programs

DURINT Topic	Agency	DURINT Team
Nanostructures for catalysis	Air Force	University of Washington, Iowa State University, University of Pittsburgh
Polymeric nanocomposites	Air Force	University of Akron
Polymeric nanophotonics and nanoelectronics	Air Force	University of Washington, University of California at Berkeley, MIT, Yale
Quantum computing with quantum devices	Air Force	Harvard University, University of Rochester
Quantum computing with quantum devices	Air Force	University of Kansas
Molecular recognition and signal transduction in biomolecular systems	DARPA/Air Force	University of Illinois at Urbana-Champaign, Harold Washington College
Synthesis and modification of nanostructure surfaces	DARPA/Air Force	University of California at Berkeley, University of California at Los Angeles, Princeton University, Louisiana State University

SOURCE: Adapted from data provided by Gernot Pomrenke, Air Force Research Laboratory.

TABLE 5-4 Nanotechnology MURIs in FY 2001

MURI Topic	Agency	DURINT Team
Multi-functional nano-engineered coatings	Air Force	University of Virginia, Ohio State University, University of Cincinnati, Arizona State University, University of New Mexico
Multi-functional nano-engineered coatings	Air Force	University of Minnesota, North Dakota State University, University of Missouri at Kansas City, University of Dayton

SOURCE: Adapted from data provided by Gernot Pomrenke, Air Force Research Laboratory.

TABLE 5-5 AFOSR Technology Grants in FY 2001

Institution	Title of Equipment Proposal	Equipment
University of California at Santa Barbara	Catalysis by Nanostructure: Methane, Ethylene Oxide, and Propylene Oxide Synthesis on Ag, Cu, or Au Nanoclusters	UHV cluster source, characterization chamber, etc.
Massachusetts Institute of Technology	Very Low Temperature Measurement System for Quantum Computation with Superconductors	Very-low-temperature-measurement system
Pennsylvania State University	Photoelectron Spectrometer and Cluster Source for the Production and Analysis of Cluster Assembled Nanoscale Materials	Photoelectron spectrometer and cluster source
University of Arizona	Nanotechnology Instrumentation	Optical parametric oscillator laser
Western Kentucky University	Acquisition of an X-ray Diffractometer for Nanotechnology Research	X-ray diffractometer
University of Virginia	Acquisition of a High-Resolution Field Emission Electron Microscope for Nanoscale Materials Research and Development	Field emission electron microscope

SOURCE: Adapted from data provided by Gernot Pomrenke, Air Force Research Laboratory.

TRENDS IN DoD AND AIR FORCE RESEARCH FUNDING

R&D funding within the Department of Defense has not kept up with R&D funding in other agencies (see Figure 5-2). Furthermore, funding of the basic research within DoD has fluctuated by about 50 percent over the past 20 years, in constant dollars. DoD's share of funding relative to the total basic research funding for all agencies has decreased by a factor of more than 2 (see Figure 5-3).[1] Since university research in engineering and the physical sciences historically has been supported largely by the DoD, this represents a significant national deemphasis on support for these important disciplines.

In addition to this decline in DoD's share of overall funding, the Air Force S&T program has suffered an even greater erosion relative to that of the other two Services over the past 11 years (Figure 5-4). Together, the slippages relative to other agencies and to the other Services mean a significant reduction in Air Force support of basic research. Over the 1989-2000 period the Air Force investment in basic research declined by $39 million in constant dollar terms, a decrease of 15 percent.[2] For the S&T budget (6.1-6.3), the integrated 46 percent decrease (in constant dollars) is even more dramatic, as Figure 5-4 illustrates.

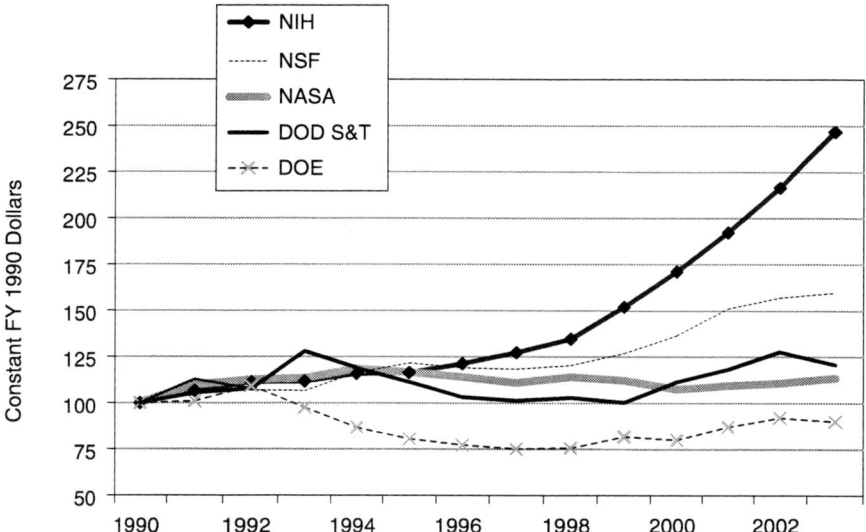

FIGURE 5-2 Trends in federal R&D funding, FY 1990-2003. SOURCE: American Association for the Advancement of Science. 2002. Trends in Federal R&D, FY 1990–2003. Available online at <http://www.aaas.org/spp/dspp/rd/cht9003a.pdf> [May 29, 2002].

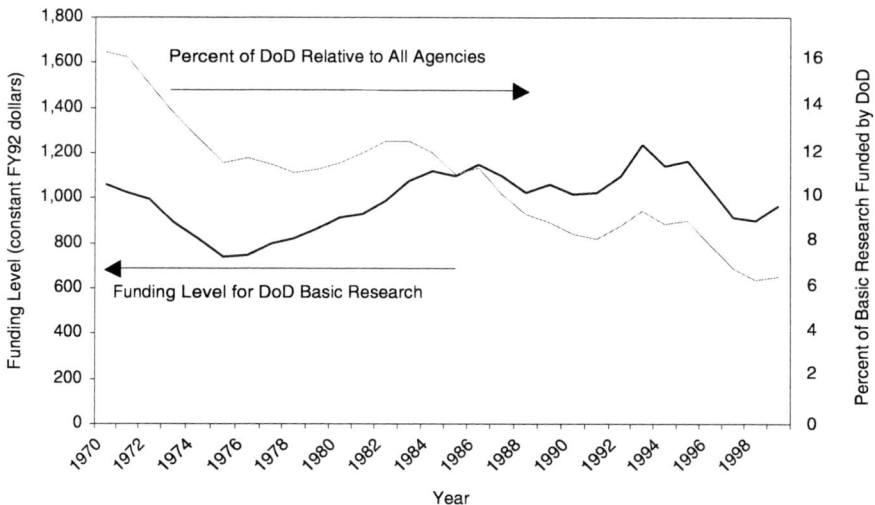

FIGURE 5-3 Funding of basic research by DoD. SOURCE: Plotted from data in National Science Foundation. 2000. Science and Engineering Indicators 2000. Available online at <http://www.nsf.gov/sbe/srs/seind00/start.htm> [May 29, 2002].

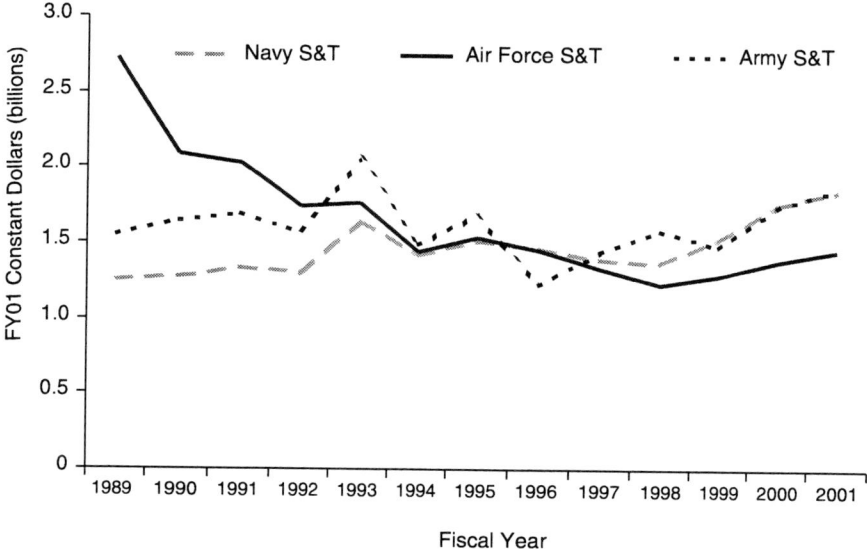

FIGURE 5-4 Science and technology funding levels by Service. SOURCE: Tuohy, R. 1999. Review of Department of Defense Air and Space Science and Technology Program. Briefing by Robert Tuohy, director, DoD Science and Technology Plans and Programs, to the Committee on Review of the Department of Defense Air and Space Systems Science and Technology Program, Wyndham Bristol Hotel, Washington, D.C., December 16.

This problem was highlighted in two recent studies:

The paucity of S&T funding in the last decade has eroded traditional Air Force technology strengths like electronic warfare. At the same time, industry basic research has shrunk dramatically, with a much shorter time horizon than 20-30 years ago.[3]

The committee believes that the reductions made by the Air Force to its S&T investment since the end of the Cold War did not take into account the changing nature of the global threat and the S&T challenges it presents. . . . The committee believes that the Air Force's current (FY01) investments in air, space, and information systems S&T are too low to meet the challenges being presented by new and emerging threats.[4]

This environment in which the Air Force finds itself cries out for increased investment in R&D. Potential nation-state adversaries as well as supranational terrorist groups connected not by national attachments but by ideology continue to pursue ways to gain advantage, raising challenges ranging from chemical and biological agent attacks, to camouflage under foliage, to deeply buried targets, to urban warfare, to new unconventional threats to civilian populations. To make progress against these difficult problems, the Air Force must continue to make well-placed, long-term investments in basic research appropriate to its missions.

These shortfalls have been recognized in congressional hearings, as noted by the House Committee on Science in its press release on the results of a 2001 NRC report:[5]

> The panel recommended that overall Air Force spending on science and technology should be increased by one-and-a-half to two times its current level.[6]

They have also been mentioned in government data analyses:

> While DoD continues to be the largest Federal funder of R&D, the FY 2002 budget request for DoD was below actual obligations in FY 1990. The DoD share of Federal obligations for R&D and R&D plant has fallen from 57 percent in FY 1990 to an expected 40 percent in FY 2002. The last time the DoD share was this small was in FY 1979 when the agency provided 43 percent of the Federal total. DoD's R&D and R&D plant dollars have dropped at an average annual rate of nearly 1 percent (a 3-percent decrease in constant 1996 dollars) between FYs 1990 and 2002.[7]

The political and economic context of DoD S&T activities has changed substantially. Industry must deal with global competition, and the commercial market has become much larger than the military market. This has several implications for DoD research:

- The commercial sector is driving near-term advances in components and systems,
- The DoD must incorporate commercial products into its systems and provide ruggedization and military specialization as an overlayer,
- Increased international competition is forcing industry to focus on the near term at the cost of longer-term investments in fundamental and applied research.

Figure 5-5 illustrates these points for semiconductor integrated circuits (ICs), a critical area of micro- and nanotechnology. Similar relationships exist in many other technology areas relevant to the Air Force, such as wireless communications and jet engines. In 1976, military purchases accounted for 17 percent of IC worldwide sales ($700 million out of total sales of $4.2 billion)—a significant market share that gave the DoD leverage in defining product specifications and directions. Over the ensuing 20 years, the military market increased only marginally, to $1.1 billion, while the commercial market exploded, to $160 billion. Today the commercial market is over $200 billion, with the military market accounting for less than 1 percent of sales. (The data for 2000 were from a different source and include all government purchases of ICs, not only military purchases.) Clearly, the commercial market has become the dominant force in setting product directions. DoD must use predominantly commercial products—

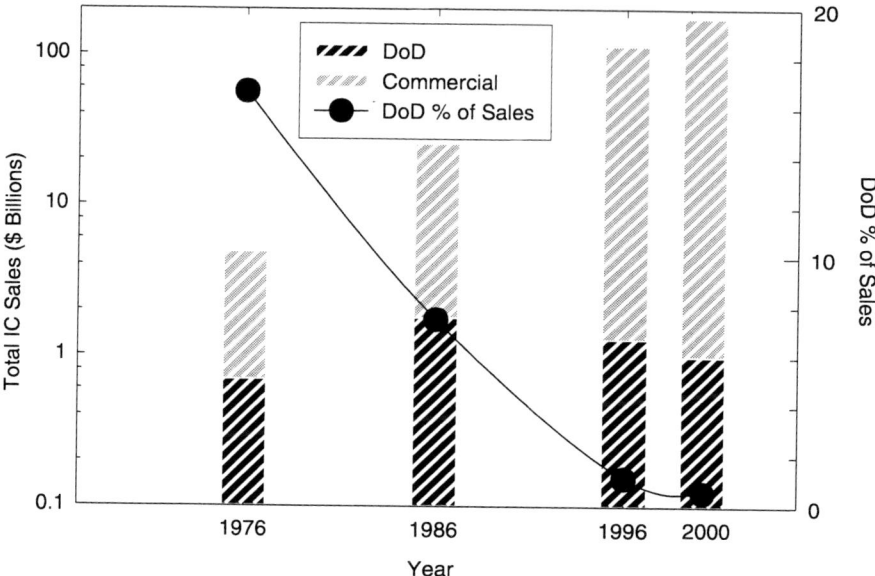

FIGURE 5-5 Integrated circuit sales. SOURCE: For 1976, 1986, and 1996 data points—National Research Council. 1999. Reducing the Logistics Burden for the Army After Next: Doing More with Less. Washington, D.C.: National Academy Press; for 2000 data point—personal communication between Brian Matas, IC Insights, and Steve Brueck, Committee on the Implications of Emerging Micro and Nano Technology, March 26, 2002.

it is no longer in a position to insist on specifications that are inconsistent with the marketplace, even if researched with DoD funds.

At the same time, competitive pressures are shortening industry development horizons to only the next one or two product cycles:

> Much of current industrial research has a very short time horizon and, in addition, tends to be focused on incremental improvements of current civilian products.[8]

In a cycle that spans S&T through development and production to life-cycle support, a few additional dollars added to industry's profit-driven, near-term development will have relatively little influence. DoD can exert much more influence by leveraging its dollars either at the front end by enabling the S&T phase or at the back end by (1) adapting commercial technology to meet DoD needs and (2) designing and procuring systems specific to DoD requirements.

Commercial industry remains dependent on the federal government for investments in long-term, higher-risk research. Historically, DoD has provided almost half of the total federal investment.[9] In FY2002, estimated expenditures by all of the Services account for about $11 billion in basic and applied research and advanced technology development (6.1-6.3).[10] The amount represents more than 36 percent of all federal investment in engineering.[11] The concentration in information technology hardware is undoubtedly much greater. The worrisome trends in DoD and Air Force funding noted above therefore imply an even greater percentage reduction in funding in these vital areas that will impact progress for both DoD and commercial applications.

U.S. industry has become increasingly dependent on government-funded research conducted at universities and federal laboratories. DoD's research investment has remained relatively stable in constant dollars because its relationship with industry has served the needs of all parties—the DoD, the research community, and the commercial sector. The imperatives of national defense have enabled DoD to invest directly in important new ideas, providing adequate resources to make a difference. Although this focused approach has not led to success in every project, it has nonetheless fostered enormous progress in new technologies for both the military and the commercial sectors, with the early stages of new technologies almost always being applicable to both sectors. Examples include devices and systems such as microwave sources, the transistor, the laser, fiber optics, and the many new materials used in aircraft technologies. By investing wisely, DoD can influence the directions of research and subsequent development and guarantee an adequate research base for technologies to meet its future needs.

DoD can also leverage its investments by coordinating teams drawn from federal laboratories, industry, and universities for demonstration projects to develop new defense applications based as much as possible on commercially available hardware and software. The advantage of U.S. forces will increasingly depend on adaptations of commercial products with a defense overlay, rather than on expensive military-specific products.

AIR FORCE INVESTMENT STRATEGY AND CHALLENGES

Micro- and nanotechnologies extend over a large area that will present many opportunities for Air Force missions. At the same time, the vast expanse and diversity of these technologies presents a significant challenge—namely, maintaining sufficient coverage in newly advancing areas while focusing investments to impact the most critical long-term missions. In this section the committee discusses the general issue of devising and maintaining the Air Force investment strategy for micro- and nanotechnologies.

There are many challenges for the Air Force. First, which of the many nanotechnologies should be supported, given a limited budget, and how can the Air Force's return on investment be maximized? Second, how should those efforts be integrated into the national nanotechnology program and how can foreign nanotechnology developments be leveraged? Third, as these micro- and nanotechnologies approach maturity, how should they be used in hardware and systems? There is a long and wide gap between discovery of a technology and its commercialization to the point where it can be incorporated into military hardware (Chapter 4). And where the Air Force undertakes to develop an item or system in the absence of commercialization, there are large risks—the Air Force cannot and should not expect success except in a minority of such projects.

The Air Force needs a cadre of dedicated professional personnel who follow nanotechnology developments. There are areas of nanotechnology in which commercial interest is very small—for example, certain sensors—and where the Air Force must act alone or in conjunction with the other Services. This cadre of researchers must be sustained by letting them have their own research projects.

For many areas, the long-term challenge is sustaining efforts at universities and other research organizations. The Air Force needs to maintain long-term funding relationships with relevant organizations and scientific and technological contributors and to use its funding leverage to ensure relevance to Air Force requirements. Lastly, communication is required between these efforts and the military and industrial organizations that plan and evolve new military systems.

To exploit these opportunities in the face of limited resources, three distinct investment strategies are needed:

- *Long term.* Long time horizons and investment strategies must be maintained for AFOSR (6.1) funding. AFOSR needs to be a leading participant in setting the long-term Air Force vision in the area of micro- and nanotechnologies, and AFRL upper-level management needs to vigorously develop and communicate its vision to the working level. Support of novel concepts and emerging areas at universities is critical, not only to maintain contact with the most creative researchers, but also to engage the best students in areas strategic for the Air Force and to cultivate them for the future. Here the emphasis should be on innovative concepts that might lead to entirely new military capabilities. In addition AFRL scientists must be active participants in the wider research community so as to take part in and react to new developments.
- *Medium term.* For the medium term there must be significant reliance on the commercial sector to explore the possibilities for implementing new concepts. Industry's large investment in 6.2-6.4-like activities for these technologies can help determine which approaches will lead to commercial application. Major areas of new development provide technologies for use in military platform systems and help to establish an understanding of their potential for manufacturability, reliability, and cost reduction. In-house experts should couple to and track these new developments to understand their potential for meeting Air Force needs.

- *Near term.* For the nearer term (6.3) investments the focus needs to be on overlaying military needs on commercial products. An example might be the use in space radiation environments of a MEMS-based microtechnology proven to have high reliability in terrestrial applications. By taking a leadership position in such targeted areas, the advances and investments of the commercial sector can be exploited and the technologies adapted to the specialized needs of Air Force missions. The committee anticipates there will be significant opportunities for this strategy in the area of micro- and nanotechnologies. The Small Business Innovative Research (SBIR) and Small Business Technology Transfer (STTR) programs provide a mechanism for involving small businesses and universities in specific near-term development needs.

Several points worthy of increased attention are the following:

- AFRL research staff need to be full members of the nano and micro research community. This will require
  —maintaining some flexible funding to enable the sustained exploratory activity necessary for participation in the community and
  —increased coupling to academia, other government laboratories, and industry researchers.
- Better in-house and external mechanisms for selection of micro- and nanotechnology efforts are needed, and realistic goals need to be set. Almost universally, the AFRL presentations did not do an adequate job of benchmarking AFRL efforts against those of the broader community. Two types of in-house experts should be nurtured—evaluators who understand emerging S&T areas and implementers who know the state of the art and can serve as smart buyers.
- Ways should be devised to attract more of the top students in micro- and nanotechnologies to AFRL—for example, by increasing graduate student involvement in AFRL projects and by increasing coupling of AFRL and university programs.

FINDINGS AND RECOMMENDATIONS

Finding P1. *Both overall DoD and—even more—Air Force policies have de-emphasized R&D spending to the detriment of DoD and Air Force long-term needs. The Air Force relies heavily on the technological sophistication of its platforms, systems, and weapons. Its ability to meet its long-term objectives is critically dependent on a strong and continuing commitment to R&D.*

Recommendation P1. *The Air Force must significantly increase its R&D funding levels if it is to have a meaningful role in the development of micro-*

and nanotechnology and if it is to be effective in harnessing these technologies for future Air Force systems. DoD and the Air Force have historically funded a majority share of the nation's research in information technologies. Their funding retrenchment represents a national de-emphasis on the future of this critically important war-fighting capability.

Finding P2. *The military market for many micro- and nanotechnologies (e.g., advanced computing, communications, and sensing) is small in comparison with commercial markets. Yet, the Air Force and DoD have mission-specific requirements not satisfied by the commercial market. Military-specific applications will not be supported by industry without government and Air Force investment, particularly in basic research.*

Recommendation P2. *The Air Force should concentrate its efforts in micro- and nanotechnology on basic research at the front end and on Air Force-specific applications at the back end.* Rather than competing with the commercial sector, the Air Force should stay strongly connected to commercial advances and adapt them to Air-Force-specific requirements.

Finding P3. *The Air Force has recognized the importance of micro- and nanotechnologies for its future capabilities and has begun a planning process to maximize the benefits of its in-house and extramural research programs. Strong leadership will be necessary to ensure maximum benefit from the Air Force Research Laboratory research programs.* The Air Force has coupled its programs with other research programs within DoD, especially those of DARPA and DDR&E.

Recommendation P3. *It will be critical to continue the planning for micro- and nanotechnologies at the highest levels of the Air Force Research Laboratory (AFRL). AFRL should also strengthen its external review processes to assist the leadership and to ensure that its work is well coordinated with national efforts. The Air Force should coordinate its initiatives with other federal agencies and work to build collaborative programs where appropriate.*

Finding P4. *The committee perceived a lack of consistency in the quality of current in-house Air Force programs and in the benchmarking of those programs against the large number of programs under way throughout the world.*

Recommendation P4. *Considering that micro- and nanotechnology is a new and rapidly emerging, interdisciplinary field, the Air Force should critically evaluate its efforts in micro- and nanotechnology to select areas of strong potential payoff for Air Force missions and to sustain the highest-quality program.* This will require the following:

- Long-term professional participation of Air Force personnel as active partners with the external micro- and nanotechnology community.
- Strong leadership and technical evaluation at the highest levels of AFRL technical leadership, as has already begun.
- Both fundamental research and focused, interdisciplinary development efforts. Fundamental research efforts are required to sustain a cadre of scientists with a deep understanding of both micro- and nanotechnology developments and of Air Force requirements. Interdisciplinary development efforts put an essential Air-Force-specific overlay on this fundamental research and force multidisciplinary teams to confront real system- and subsystem-level problems, which is essential for bringing any technology from the laboratory bench to practical application.

REFERENCES

1. National Science Foundation. 2000. Science and Engineering Indicators 2000. Available online at <http://www.nsf.gov/sbe/srs/seind00/start.htm> [July 9, 2002].
2. National Research Council. 2001. Review of the U.S. Department of Defense Air, Space, and Supporting Information Systems Science and Technology Program. Washington, D.C.: National Academy Press.
3. Air Force Association. 2000. Shortchanging the Future January: Air Force Research and Development Demands Investment. Arlington, Va.: Air Force Association.
4. National Research Council. 2001. Review of the U.S. Department of Defense Air, Space, and Supporting Information Systems Science and Technology Program. Washington, D.C.: National Academy Press.
5. National Research Council. 2001. Review of the U.S. Department of Defense Air, Space, and Supporting Information Systems Science and Technology Program. Washington, D.C.: National Academy Press.
6. U.S. House of Representatives. 2001. Blue-Ribbon Panel Warns of Dangers of Reduced Investment in Defense Science and Technology, Committee on Science Press Release, July 27. Available online at <http://www.house.gov/science/press/107pr/107-66.htm> [July 10, 2002].
7. Meeks, R.L. 2002. Changing Composition of Federal Funding for Research and Development and R&D Plant Since 1990, National Science Foundation InfoBrief NSF 02-315, April. Available online at <http://www.nsf.gov/sbe/srs/infbrief/nsf02315/nsf02315.pdf> [July 10, 2002].
8. U.S. House of Representatives. 2001. Pentagon Advisory Panel Criticizes Science Budget, Press Release, March 19. Available online at <http://www.house.gov/tonyhall/pr214.html> [July 10, 2002].
9. American Association for the Advancement of Science. 2002. Trends in Federal R&D, FY 1990–2003. Available online at <http://www.aaas.org/spp/dspp/rd/cht9003a.pdf> [July 10, 2002].
10. U.S. House of Representatives. 2001. Pentagon Advisory Panel Criticizes Science Budget, Press Release, March 19. Available online at <http://www.house.gov/tonyhall/pr214.html> [July 10, 2002].
11. U.S. House of Representatives. 2001. Pentagon Advisory Panel Criticizes Science Budget, Press Release, March 19. Available online at <http://www.house.gov/tonyhall/pr214.html> [July 10, 2002].

6

Opportunities in Micro- and Nanotechnologies

OVERARCHING THEMES

In this chapter the committee explores the implications of the emerging micro and nanotechnologies for future Air Force systems applications. As discussed in Chapter 5, four overarching themes emerge from this study of micro and nano technologies:

- increased information capabilities
- miniaturization of systems
- new materials resulting from new science at these scales
- increased functionality and autonomy

These trends are pervasive throughout the advances in micro- and nanotechnologies, and the new capabilities they provide will have far-reaching consequences for Air Force missions. The committee now discusses the implications of each of these trends.

Increased Information Capabilities

The committee sees a continued scaling of microelectronic, magnetic, and optical devices to smaller size and higher densities at ever-higher speeds. The result will be the ability to store, process, and communicate an ever-increasing amount of information. Quoting from *Joint Vision 2020*:

> Information, information processing, and communications networks are at the core of every military activity. Throughout history, military leaders have re-

garded information superiority as a key enabler of victory. However, the ongoing "information revolution" is creating not only a quantitative, but a qualitative change in the information environment that by 2020 will result in profound changes in the conduct of military operations. In fact, advances in information capabilities are proceeding so rapidly that there is a risk of outstripping our ability to capture ideas, formulate operational concepts, and develop the capacity to assess results. While the goal of achieving information superiority will not change, the nature, scope, and "rules" of the quest are changing radically.[1]

From the analysis in Chapter 3 it is apparent that the current rapid increase in the ability to handle information will continue at least for the next decade and beyond. The doubling of computing power every 18 months and the even more rapid increase in information transmission rate and storage capacity will lead to an increase of at least 128× in the amount of information that can be gathered and processed. Today's smart weapons will seem "mentally challenged" 10 years from now.

The trend in information density is important because so many information technologies are foreseen to have a significant impact on future military operations. Examples include these:[2]

- Autonomous and adaptive algorithms for resource scheduling, mission planning, and mission execution
- Artificial/virtual intelligence (AI/VI), self-awareness, intuitiveness, automated recognition
- Human-machine interfaces and robotics
- Heterogeneous databases, software, integration, modeling and processing techniques
- Advanced tools and algorithms for modeling and simulation (M&S)
- Satellite onboard data processing and storage
- Nonvolatile random access memory
- Mass storage memory (including optical storage technologies)
- Radiation hardening and shielding of components
- Plug-and-play hardware and software technologies

Beyond this relatively near-term trend, the committee anticipates that emerging nanotechnologies will enable even more revolutionary long-term changes in how we obtain and use information. Exploiting these advances will be an important and challenging task for the Air Force.

Miniaturization

The reduction in size of systems from computers to cell phones is a continuing evolution for electronic systems. The significance of this miniaturization goes well beyond just the smaller size and reduced weight. Batch fabrication, the parallel manufacturing of many integrated components, has been a key driver in the miniaturization of microelectronics because it reduces cost and increases reliability. Another significant trend is the integration of components and subsystems into fewer and fewer chips, enabling increased functionality in ever-smaller packages. These trends are extending to include microelectromechanical systems (MEMS) and other technologies for sensors and actuators, thus allowing the possibility of miniaturizing entire systems and platforms. The combination of reduced size, weight, and cost per unit function has significant implications for Air Force missions, from global reach to situational awareness. Examples may include the rapid low-cost global deployment of sensors, launch-on-demand tactical satellites, distributed sensor networks, and affordable UAVs.

New Engineered Materials

Advances in micro- and nanofabrication technologies are enabling the engineering of materials down to the atomic level. While design and fabrication capabilities are still primitive from an applications perspective, there is great potential for improving the properties and functionality of materials. Examples of recent advances in materials range from carbon nanotubes with great strength and novel electronic properties, to quantum dot communication lasers, to giant magnetoresistive materials for high-density magnetic memories. Theory and simulation will play an increasingly important role in guiding the development of new nanostructured materials and of systems based on such materials. By combining materials at the micro- and nanoscales to form smart composite structures, additional increases in functionality can be achieved. New materials are an underlying enabling capability. They will be used to expand the performance envelope of electronics, sensors, communications systems, avionics, air and space frames, and propulsion systems. Theory and modeling of materials are advancing significantly, as is our understanding of the relationships between material composition, properties, and structure. Over time, these advances may reduce the long lead time for developing new materials and may also help in the design of new, more functional materials with impact on Air Force systems.

Increased Autonomy and Functionality

The advances in information density, miniaturization, and materials functionality will enable a degree of autonomous operation for systems that cannot be fully envisioned today. Enhanced functionality and increased autonomy based on micro- and nanotechnologies have many systems benefits:

- lower risk for humans
- higher performance
- lower cost platforms

- reduced communication requirements with a correspondingly lower probability of detection

Initially, this autonomy will be seen simply as an evolutionary extension of the capabilities of current systems such as cruise missiles or UCAVs, providing increased accuracy and range or other performance advantages. Over the longer term, however, the dramatic increases in local information awareness and computational power will enable independent decision making and will have a dramatic impact on the conduct of warfare. Systems may also be able to power, self-repair, and reconfigure themselves to extend the scope of their missions. The lowered cost and increased functionality will lead to swarms of intelligent agents with emergent behavior that differs from that of any single entity. Integrating these advances into the Air Force concept of operations (CONOPS) will be challenging and will raise important global political and societal issues as well, such as the acceptable bounds of future warfare—for example, specifying the roles of autonomous decision-making machines in war fighting.

AIR FORCE MISSIONS AS DRIVERS FOR MICRO- AND NANOTECHNOLOGIES

Micro- and nanotechnology's potential for reducing weight and size while enhancing performance has particular relevance to the Air Force mission of defending the United States through control and exploitation of air and space. The benefits of significant miniaturization, reduced cost, and increasing performance, if they can be achieved, would be particularly significant. The advance of information technologies provides the clearest demonstration of this promise (one place where it was especially influential is avionics). Emerging areas such as MEMS and integrated sensor systems point the way to new opportunities. Miniaturization of systems, combined with batch fabrication and integration of components, may lead to significant improvements in affordability, enabling more densely distributed systems. Further, improved performance of materials through nanostructuring and other advances in nanoscience and technology could enable systems opportunities of a more revolutionary nature, so that such advances merit careful tracking by the Air Force.

Future opportunities of micro- and nanotechnologies are relevant to all six of the core competencies in the Air Force strategic plan:

- aerospace superiority
- information superiority
- global attack
- precision engagement
- rapid global mobility
- agile combat support

Increased functionality in smaller packages will enable greater reach, more precise engagement, greater global awareness and knowledge, and more agility in the control of air and space. Understanding how the capabilities of advancing micro- and nanotechnologies can be exploited in new systems that will enhance missions is a major challenge worthy of significant attention when planning. Changes are likely to be required not only in specific platforms but also in the organization of the Air Force to fully exploit the game-changing advances that are likely to flow from the application of micro- and nanotechnologies to its missions. These are well beyond the scope of the present report but should not be ignored as these technologies advance into systems and platforms.

The Air Force will be challenged to use the advances in micro- and nanotechnologies in its hardware and systems. There is a long and wide gap between the discovery of a technology and its commercialization to the point where it can be incorporated into military hardware. Coordination of S&T investments while considering both Air Force mission planning and external advances in science, technology, and commercial development will be critical. Although it is impossible to predict with certainty the systems implications, mission impact, and commercial viability of these emerging advances, it is important to consider their possible benefits. The continually changing competitive environment with respect to threats from adversaries underlines the need for alternative approaches to some of the most daunting threats. Based on the technologies discussed in Chapters 3 and 4, the committee now discusses some specific opportunities that could emerge from Air Force investments in micro- and nanotechnology S&T in these areas.

AREAS OF OPPORTUNITY

In this section the committee identifies selected opportunities for linking micro- and nanotechnologies to Air Force mission platforms. While pursuit of any of the specific systems opportunities depends on considerations well beyond the scope of the present study, it would be beneficial to examine these or related opportunities to determine their fit with Air Force needs. As appropriate, key issues should be further examined through focused S&T projects within the AFRL, support to the broader R&D community through focused AFOSR grants and AF contracts, and related program strategies.

An overview of several opportunity areas is given in Table 6-1, which couples the opportunities to the micro- and nanotechnology S&T areas discussed in Chapters 3 and 4 (left side). Also shown is the relationship of these opportunities to the Air Force core competencies (right side). The approximate time frame in which these technologies could have an impact on Air Force missions is near term (0-10 years), medium term (10-20 years), or long term (20-50 years). The technologies are grouped by use: space vehicles and systems, weapon systems, and air vehicles and systems.

Space Vehicles and Systems

The miniaturization of systems enabled by micro- and nanotechnologies while maintaining or increasing capabilities provides an important opportunity for space systems. These advances will enable satellites to carry out more functions consistent with increased information superiority and to employ lighter vehicles or arrays of satellites, providing new options for launch vehicles and for distributed sensing or communications from space. Because of their reduced weight and cost, nano- or picosatellites may enable new functions (see Box 6-1 for a current example).

At the same time, adopting the manufacturing advances pioneered in the microelectronics industry—batch processing and increased quality control—along with anticipated future advances in nanotechnologies such as self-repair and reconfigurability will significantly expand the range of possible applications. Low-power systems in combination with more efficient space power and energy storage systems will enable long lifetimes. Future ultrahigh-resolution sensor systems will be enabled by microarray technology for imaging sensors, on-board digital image processing, and wide band communications. Large arrays, made possible by miniaturization and the associated reduced weight and cost, and phased arrays will enable continuous surveillance. The ability to monitor with high spatial resolution over essentially all wavelengths will allow truly realizing continuous total information systems. Specific areas suggested for further consideration are distributed satellites, integrated spacecraft, and micro-launch vehicles.

Distributed Satellites (Medium to Long Term)

Two or more satellites with suitable coherent signal combining can have the resolving power of a much larger spacecraft. The angular resolving capability is roughly equal to λ/D, where λ is wavelength (about 0.5 micrometers for visible light and anywhere from millimeters to a meter for radio frequencies) and D is the separation between spacecraft. Distributed spacecraft are the only practical way to generate milliradian-wide beam widths in space at UHF frequencies (300-1,000 MHz); the required kilometer-scale antenna diameters are impossible using a single antenna structure. Possible applications include battlefield cellular communications systems, where the cells are generated by extremely narrow spot beams from satellites, and geolocation of UHF jammers. Geolocation can be accomplished using two spacecraft, while synthesis of a narrow beam antenna for communications will require hundreds of individual spacecraft swarms in a sparse aperture array. Mass-produced micro-, nano-, and picosatellites with orbit and attitude control capability make the latter scenario possible. The key technology development that is still needed for such arrays to work is handling the phase of the signal from this multitude of separate transmitters and receivers. Technology

TABLE 6-1 Selected Mission and Platform Opportunity Areas

System Type	Science and Technology Area						Selected Mission and Platform Opportunities	Time Scale[a]	Air Force Critical Future Capability					
	Information Technology	Sensors	Bioinspired Materials and Systems	Structural Materials	Aerodynamics, Propulsion, and Power				Aerospace Superiority	Information Superiority	Global Attack	Precision Engagement	Rapid Global Mobility	Agile Combat Support
Space vehicles and systems	X	X	X	X	X		*Distributed satellite.* Self-sustaining nano-satellite arrays/swarms to monitor, report status, and take action	M-L	X	X	X		X	X
	X	X	X	X	X		*Integrated spacecraft.* Highly integrated, reprogrammable, reconfigurable systems	M	X	X	X	X		X
				X	X		*Micro launch vehicles.* Low-cost, launch-on-demand tactical space systems	M-L	X	X	X	X	X	

OPPORTUNITIES IN MICRO- AND NANOTECHNOLOGIES 207

Category	Capability	Term[a]										
Weapon systems	*Miniaturized ballistic missiles.* Rapid global-reach system enabled by microtechnologies	M	X	X		X	X			X	X	X
	UAV-launched ABM boost-phase interceptors. Micro- and nano-enabled small missile interceptors	M	X	X		X	X			X	X		
	Air-to-air and air-to-ground weapons. Missiles and bombs with significantly reduced weight, size, and cost through miniaturization with better performance	M	X	X		X	X				X	X	X
Air vehicles and systems	*Micro air vehicles.* Low-cost, ubiquitous, autonomous surveillance and reconnaissance systems and microdecoys; cooperative behavior of swarms of vehicles	N-M-L	X	X	X	X	X			X		X	X
	MEMS-based active aerodynamic flight control. Microsensing and control of air flow combined with new materials for enhanced flight efficiency	M-L	X	X	X	X	X			X		X	X

NOTE: Also indicated is their relation to micro- and nanotechnology S&T areas, as discussed in Chapter 3, and to Air Force–defined critical future capabilities.

[a] N, near term; M, medium term; L, long term.

BOX 6-1
Nano- and Picosatellites

Satellite system and subsystem technology has evolved to the point where nanosatellites and picosatellites can perform complex scientific, communications, Earth observation, and satellite assistance missions. Great Britain's first nanosatellite, the 6.5-kilogram mass SNAP-1, was launched along with the Tsinghua-1 microsatellite on June 28, 2000, to attempt a rendezvous and on-orbit inspection of a target spacecraft (Tsinghua-1). The ~30-centimeter-scale SNAP-1 had four ultra-miniature CMOS active pixel array video cameras and a 12-channel GPS navigation system to enable autonomous orbit maneuvers using a simple 50-millinewton cold gas thruster.[1] Although SNAP-1 failed to rendezvous with Tsinghua due to differential air drag between the two spacecraft, it demonstrated that nanosatellites with autonomous navigation and propulsion systems capable of formation flying could be readily fabricated using existing technology.

Figure 6-1-1 shows a pair of 260-gram-mass picosatellites developed by the Aerospace Corporation with assistance from Rockwell Scientific under DARPA support. These 4 × 3 × 1-inch battery-powered spacecraft were orbited on Janu-

FIGURE 6-1-1 DARPA/Aerospace Corp. picosatellites. Photograph reprinted with the permission of the Aerospace Corporation.

ary 26, 2000, followed by two more on July 29, 2000. They served as inexpensive on-orbit testbeds for DARPA-sponsored MEMS RF switches and low-power RF networking technologies. Nano- and picosatellites are ideal for inexpensive, fast turnaround missions; the DARPA picosatellites were designed, fabricated, and tested within 6 months. Picosatellite development in the United States is sponsored by DARPA and AFRL under the MEMS-based Picosat Inspector program. A number of flight demonstrations with increasing sophistication (MEMS inertial navigation and propulsion) will evolve into a picosatellite inspector that can be ejected from a host satellite.

CMOS, MEMS, and related fabrication techniques have provided small, low-power sensors and imaging arrays on silicon dice that can be utilized for attitude determination. The Honeywell HMC1023 three-axis magnetic sensor, for example, can fit on a U.S. dime yet has a minimum detectable field of 85 microgauss, which is more than sufficient to provide spacecraft orientation to within a degree with respect to the Earth's local magnetic field. CMOS imagers such as the Agilent HDCS-1020 active pixel imaging chip could readily be adapted for use in an imaging Sun sensor with 352×288 pixel resolution; it could provide better than 1/3-degree resolution over a 90-degree field of view. Swapping the CMOS photodiode structure with polysilicon-aluminum micro-thermocouples, also available in the basic CMOS process, would result in a thermal imaging system that could detect the 300 K Earth against the 3 K background of space for use in an Earth horizon sensor.[2]

Nanosatellites can be fabricated using current techniques, but capable picosatellites and smaller spacecraft will require higher levels of integration. System-on-a-chip technologies, high-speed serial interfaces and networking, and increased device density due to better packaging (e.g., flip chip-on-a-board) will enable current-generation microsatellite electronics to fit within a cubic inch. Related attitude sensors, MEMS inertial sensors, and propulsion systems could fit within a similar volume for some applications, like the satellite inspector. Even smaller volumes will be possible as IC device densities continue to improve.

AFRL currently has an exploratory program with three universities (Arizona State University, the University of Colorado, and New Mexico State University) to develop and test a cluster of three next-generation, toaster-sized nanosatellites for launch from the space shuttle for operation in a distributed mode. Potential advantages for such nanosatellite distributed arrays include reduced launch costs, increased reliability (losing one satellite would not preclude use of the array), and more rapid system development cycles. The really significant surveillance game-changer would be a cluster of nanosatellites functioning in a manner that mimics an antenna miles in diameter.

---

[1]Underwood, C., V. Lappas, G. Richardson, and J. Salvignol. 2002. SNAP-1–Design, construction, launch and early operations phase results of a modular COTS-based nano-satellite. Pp. 69–77 in Smaller Satellites, Bigger Business? Concepts, Applications and Markets for Micro/Nanosatellites in a New Information World. Boston, Mass.: Kluwer Academic.

[2]Janson, S.W. 2002. Nanotechnology—Tools for the satellite world. Pp. 21–30 in Smaller Satellites, Bigger Business? Concepts, Applications and Markets for Micro/Nanosatellites in a New Information World. Boston, Mass.: Kluwer Academic.

developments in low-power, space-hardened digital and radio-frequency electronics, microthrusters, and low-mass, low-power, short-range (kilometer-scale), free-space optical transmission systems are required. Collective arrays of satellites that function in a synchronized fashion promise significant new opportunities in capabilities and robustness of satellite systems.

Integrated Spacecraft (Medium Term)

Semiconductor batch-fabrication techniques allow cost-effective production of million-transistor digital circuits, analog circuits, radio-frequency and microwave circuits, and microelectromechanical systems (MEMS). Current trends in semiconductor and MEMS fabrication, e.g., continually increasing functionality per square millimeter of silicon, argue for the development of system-on-a-chip (SOC) technology for various spacecraft systems. Increased electronic functionality, coupled with nonvolatile memory technologies such as FLASHRAM and MRAM, enable intelligent micro- and nanosystems that can be reprogrammed or reconfigured on orbit. Possible examples of spacecraft SOCs include Sun and horizon sensors, inertial measurement units (IMUs) composed of MEMS accelerometers and rate gyros, GPS receivers for navigation and attitude determination, and MEMS-based microthruster systems. SOC interconnection can use high-speed serial lines to allow plug-and-play assembly much like the USB bus used for personal computer peripherals. The challenge will be to migrate commercial technologies into radiation-hard or radiation-tolerant technologies for use in space. The benefit will be decreased parts count per spacecraft, increased functionality per unit spacecraft mass, and the ability to mass produce micro-, nano-, and picosatellites for launch-on-demand tactical applications (e.g., inspector spacecraft) and distributed space systems.

Micro Launch Vehicles (Medium Term)

Enabled by subsystems and components such as MEMS liquid rocket engines, valves, gyros, and accelerometers, microlaunch vehicles are feasible in the size range 15 to 800 kilograms gross liftoff weight (GLOW). The payloads would make extensive use of micro- and nanotechnology as well. As an example, a 170-pound GLOW, two-stage rocket (the weight of an AIM-9) could deliver one or two kilograms to low Earth orbit. Such vehicles would be about the size and complexity (less the seeker) of a small tactical missile and thus should cost about the same. Launched from the ground or the air, the micro launch vehicle redefines the concept of low-cost access to space as cost per mission rather than cost per pound of payload (which is about the same as for larger launchers). Thus, it will be possible to place a payload (albeit a small one) into orbit for $10,000 to $50,000 rather than the $10 million to $50 million required today. This cost is sufficiently low that launchers can be stockpiled for launch-on-demand access to space, encouraging routine tactical space operations. Air launch (from a tactical aircraft for example) offers the added advantages of launch site flexibility, elimination of the requirement for a launch range with its attendant costs, and covertness (a liquid-fuelled vehicle this small might be very difficult to observe from space; also, there need be no fixed launch location). Orbital applications might include visual and IR inspection of space objects, ELINT, jamming of satellites, and antisatellite operations.

Weapon Systems

Miniaturized Ballistic Missiles (Medium Term)

The same MEMS technologies (propulsion, guidance, control, etc.) required for microlaunch vehicles enable intercontinental tactical ballistic missiles with sensor or nonnuclear munitions payloads. A 170-pound GLOW, two-stage rocket (about the weight of an AIM-9) could deliver about 10 pounds to 4,500 nautical miles or 30 pounds to 1,000 nautical miles (the reentry vehicle might account for 25 to 40 percent of this weight). The advantages of the ballistic approach compared with the air-breathing cruise missiles approach are very high speed and immunity to interception. Global, rapid deployment of sensors (and even micro air vehicles) is one obvious application. Another is weapons delivery. Enhanced effectiveness warheads pack sufficient lethality into a small package that these small missiles may be effective against relatively soft targets such as radar sites and armored vehicles. Now that much of the target detection and acquisition is done off-board the shooter platform, the long range and high speed suggest that this may be an effective approach to the now-difficult problem of attacking rapidly redeployable or moving targets, such as missile launchers in Iraq. The ultralong range eliminates the "tyranny of distance" that plagues operations in places like Afghanistan, since any target in the world is within practical range. This is truly a global reach system.

UAV-Launched Antiballistic Missile Boost Phase Interceptors (Medium Term)

Air-launched boost-phase intercept schemes could greatly increase the payload and endurance capabilities of launch platforms. Ground-launched missile interceptors are now sized according to the sensor and propulsion systems, since a direct-hit, kinetic-kill vehicle need not carry a warhead. Advanced sensors, propulsion, and other subsystems enabled by micro- and nanotechnologies will permit dramatically reduced missile mass (perhaps well below 100 kilograms), with several profound system-level impacts. One impact is that microinterceptors will have unit costs at least one and possibly two orders of magnitude lower than current designs. The miniaturization could enable UAV-launched missile boost-phase interceptors. Multiple, simultaneous launches against a single target could increase reliability, reduce the performance requirement relative to that needed for target-decoy discrimination, and permit new, more flexible system architectures. This may, for the first time, shift the economics of missile offense versus defense in favor of the defender.

Air-to-Air and Air-to-Ground Weapons (Medium Term)

Advances in micro- and nanoscale technologies offer the opportunity to explore designs for air-to-air missiles and air-to-ground bombs with significantly better performance. It may be possible to reduce weight, size, and cost by miniaturizing of sensors, avionics, and inertial measurement units in both the near and the far term. Recent improvements in long-range missile accuracy have resulted in significantly reduced loss of expensive missile-carrying aircraft in combat situations. Adoption of the next generations of MEMS-based IMUs, providing further improvements in missile reliability and accuracy, will be accelerated to further improve aircraft combat survivability. Reductions in guidance system cost and package size will translate into system savings for short-range air-to-ground and air-to-air missiles. Coupled with enhanced explosives under development and further evolutionary cost and accuracy improvements in MEMS IMUs and seekers, further reductions in size and cost may be realized, allowing significantly larger numbers of missiles or smart bombs to be carried by an aircraft. This payload quantity extension can be traded for range or logistics support size and cost as overall systems requirements dictate.

Air Vehicles and Systems

Micro Air Vehicles (Near, Medium, and Long Term)

The acronym MAV generally refers to 6-inch or smaller flying platforms, which are under development for local surveillance and reconnaissance applications. Given micro- or nano-enabled or -enhanced sensors and subsystem technologies, MAVs can grow dramatically in capability, achieving autonomy and the functionality of systems that are currently many times larger, or can shrink in size to insectlike dimensions. Current MAVs fly at 10 meters per second for 5 kilometers, but future systems could fly transonically for 1,000 kilometers or endure for tens of hours (see Box 6-2 for a current example).

In addition to performing sensing, surveillance, and reconnaissance, enhanced capability systems might function as microdecoys or carry jammers (which need not radiate much power if they are very close to the receiving antenna) or even be weaponized. Replacing larger air vehicles with MAVs promises to greatly reduce cost. The affordability of MAVs could allow large swarms of vehicles with cooperative behavior and could herald new and more robust systems approaches to reconnaissance and surveillance. Also, the lower overall

BOX 6-2
The Black Widow Micro Air Vehicle

Figure 6-2-1 shows a 6-inch wingspan, 80-gram-mass Black Widow fabricated by AeroVironment of Monrovia, California. Funded by DARPA's Tactical Technology Office, these first-generation MAVs are intended to provide local reconnaissance for soldiers at the platoon level.[1] Air Force applications could include local reconnaissance and communications relay for downed pilots, delivery of microwarheads (10 grams of explosive) with ultrahigh precision, and delivery of covert unattended ground sensors. Their small size (less than 15 centimeters, in any dimension) gives MAVs a degree of stealth.

FIGURE 6-2-1 The AeroVironment Black Widow micro air vehicle. SOURCE: Grasmeyer, J.M., and M.T. Keennon. 2001. Development of the Black Widow Micro Air Vehicle, AIAA Paper 2001-0127. Reston, Va.: American Institute of Aeronautics and Astronautics, Inc.

---

[1]McMichael, J.M., and M.S. Francis. 1997. Micro Air Vehicles–Toward a New Dimension in Flight. Available online at <http://www.darpa.mil/tto/MAV/mav_auvsi.html> [July 10, 2002].

continues

Commercially available micro- and nanotechnology in the form of two microprocessors, stamp-size 433-megahertz command receivers, and 2.4 gigahertz data transmitters, two-axis magnetic attitude sensors, MEMS pressure sensors, and a piezoelectric rate gyro make the avionics suite for this 6-inch MAV possible. Figure 6-2-2 shows the relative sizes and masses of these systems with respect to the airframe. Primary mass drivers are the batteries, the electric motor, and the power and propulsion system. Primary power loads are the electric motor (about 5 watts) and the video transmitter (0.55 watts). Microturbojets with ~10 grams of thrust would offer at least an order-of-magnitude increase in flight time from the current limit of ~30 minutes.

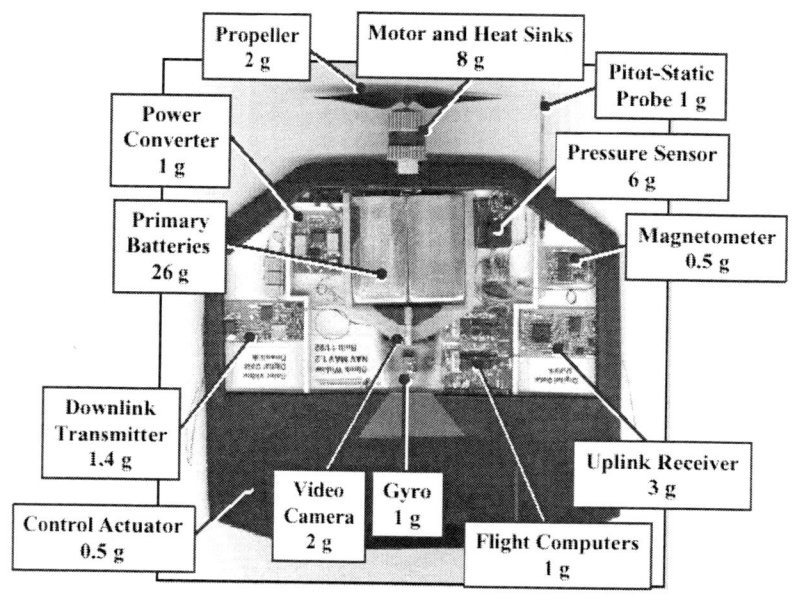

FIGURE 6-2-2 Subsystem layout, size, and mass of the Black Widow. SOURCE: Wilson, S.B. 2000. Palm Power Workshop for Micro Air Vehicles, November 15. Available online at <http://www.darpa.mil/dso/thrust/md/palmpower/presentations/wilson_part1.pdf> [April 8, 2002].

mass of a system-level solution implies a dramatically reduced logistics tail. Low cost might be traded for area coverage (by using more vehicles) or operation in very high risk environments. Large numbers flying under the tree canopy or perching in cities may be one solution to finding hidden targets.

MEMS-Based Active Aerodynamic Flight Control (Medium to Long Term)

Microsensing and control of airflow over vehicle surfaces, combined with new ultrastrong, lightweight materials, could lead to a new generation of aircraft providing enhanced flight efficiency and maneuverability without conventional rudders or other macroscopic control surfaces. These MEMS-based active aerodynamic flight control vehicles (MACs) could exploit advances in microscale sensors and actuators in combination with information technologies to provide local feedback control. Vehicle surfaces would rapidly sense and change airflow boundary layer conditions, a capability now possible with micromechanical devices and continuously increasing computing power. Such control strategies might reduce, on average, the turbulent nature of aerodynamic flow, leading to laminar flow vehicles with dramatically greater range-payload capabilities than those of current aircraft. The implications for air combat support, global reach, and reduced overseas footprint could be significant. A second aspect would be the ability to manipulate boundary layers to generate large forces and moments for flight control, possibly supplementing or replacing large-scale control surfaces while reducing weight and increasing maneuverability. The possible mission implications of MACs would appear to be worth exploring in concert with the advancement of micro- and nanotechnologies.

FINDING AND RECOMMENDATION

Finding T8. Four overarching themes emerge from the advance of micro- and nanotechnologies—increased information capabilities, miniaturization, new engineered materials, and increased functionality/autonomy. These themes could have a significant military impact by enabling new systems approaches to Air Force missions.

Recommendation T8. The Air Force should continue to study new systems opportunities that may emerge from the successful development of micro- and nanotechnologies and use these studies to help focus its applied research and development investments in these technologies.

REFERENCES

1. Joint Chiefs of Staff. 2000. Joint Vision 2020. Washington, D.C.: Government Printing Office.
2. Office of the Secretary of Defense. 2000. Space Technology Guide FY 2000–2001. Washington, D.C.: Office of the Secretary of Defense, Assistant Secretary of Defense (Command, Control, Communications, and Intelligence); Director, Defense Research and Engineering.

7

Findings and Recommendations

The committee reached a number of critically important broadly applicable findings and recommendations, which are presented in the first section of this chapter. Findings and recommendations related to technological developments in micro- and nanotechnologies are followed by policy recommendations. Findings and recommendations are presented in a logical flow. The numbering does not represent a rank ordering but simply serves as an identifier. In addition, a number of the committee's more specific findings and recommendations are listed in the second section of the chapter.

CRITICAL FINDINGS AND RECOMMENDATIONS

Technology

Four overarching themes emerged from the committee's study of the implications of emerging micro- and nanotechnologies: increased information capabilities, miniaturization, new materials, and increased functionality and autonomy (T8). The following findings and recommendations attempt to capture the essence of these themes with some specificity. The increased information capabilities flow from near-term continuation of the scaling of silicon electronics (T1) and from new and alternative concepts arising from nanotechnology research (T2). Biological science, both as inspiration (biomimetics) and as a functional contributor, offers new opportunities (T3) that build on and complement traditional sensing, computing, and communications approaches. Increased information capabilities and miniaturization together will make possible large distributed arrays of sensors (sensor swarms) on combinations of fixed and movable platforms. These array systems will exhibit new or emergent properties significantly different from those of individual components and will allow increasingly autonomous operation of Air Force systems (T4). Harnessing the capabilities of microelectromechanical systems (MEMS) to propulsion and aerodynamics will allow miniaturization of air and space platforms (T5). Maximizing the utility of these growing capabilities in information technology, biomimetics, individual sensors and sensor swarms, and MEMS actuators for the Air Force will demand specific attention to system design, architecture, and software for system implementation (T6). Because of the wide range of new capabilities being enabled, the trend to merging heterogeneous materials systems and to expanding the range of materials in micro- and nanoscale devices and systems is inexorable (T7).

Finding T1. *Further miniaturization of digital electronics with increased density (~128×) is projected by the integrated circuit industry over the next 15 years based on continued scaling of current technology.* The most recent ITRS forecasts the accelerated introduction of smaller dimensions and greater computational power than were forecast by the ITRS 2 years ago.

Recommendation T1. *The Air Force should position itself to take advantage of the advances predicted by the Information Technology Roadmap for Semiconductors.* Dramatic advances are predicted for device technology. Software, application-specific integrated circuits (ASICs), embedded computers integrating software and hardware for specialized applications, and radiation-hardening and packaging for hostile environments must be designed by, and for, the military, to take advantage of these advances.

Finding T2. *In anticipation of an ultimate end to the historical scaling of today's integrated circuit technology, many new and alternative concepts involving nanometer-dimensioned structures are being examined.* As yet, none of these concepts had demonstrated the necessary functionality and integrability to be a clear choice for "beyond silicon." Many different material and device technologies will need to be explored well into the future. Two facts seem clear. First, it is not possible to make reliable, long-term predictions of breakthrough capabilities emerging from the rich frontier of discovery, fabrication, and material properties at nanometer dimensions. The numerous avenues of research investigation are likely to uncover unexpected processes and/or material properties that will have an impact at the fundamental level of information processing. Second, it seems likely that the initial applications of any of these technologies will build on and enhance the very strong base of existing integrated circuit technology, which will provide the necessary backbone of functionality and integrability until an entirely new computation paradigm emerges.

Recommendation T2. *Exploration of the scientific frontiers involving new procedures for fabrication at nanodimensions and new nanoscale materials, properties, and phenomena should be supported.* The Air Force should track, assimilate, and exploit the basic ideas emerging from the research community and continue to support both intra- and extramural activities. The focus should be on understanding the fundamental processes for fabrication, and on the unique properties of materials and devices structures at nanometer dimensions. Extremely dense arrays of devices capable of manipulating bits rapidly and reliably should be a dominant aspect of these investigations. Individual devices with nanometer or molecular dimensions are demonstrating logical functions on a small scale with a limited number of examples. Molecular electronics appears promising at present. There is potential for new device innovations and for progress in computing architectures and strategies. Quantum computing and quantum cryptography are examples of the applications that may be enabled by further progress in micro- and nanotechnology. The technology may develop rapidly once the scientific principles and technological advantages are discovered and understood.

Finding T3. *Biological science offers new opportunities in nanotechnology systems, especially for sensors, materials, communications, computing, intelligent systems, human performance, and self-reliance.* Millions of years of evolution have produced highly specialized sensing and communication capabilities in nature. Understanding of how these sensors work is growing but is still very limited. As the fundamental mechanisms are discovered and studied, applications rapidly follow. Advances in micro- and nanotechnology have enabled discovery in biological systems, which in turn has provided new means of sensing and communicating. Clearly, advances in technology and in the biological sciences go hand in hand in developing new capabilities.

Recommendation T3. *The Air Force should closely monitor the biological sciences for new discoveries and selectively invest in those that show a potential for making revolutionary advances or realizing new capabilities in Air Force-specific areas.*

Finding T4. *Large, distributed fixed arrays and moving swarms of multispectral, multifunctional sensors will be made possible by emerging micro- and nanotechnology, and these will lead to significant fundamental changes in sensing architectures.* Concepts such as smart dust and distributed communication networks actively exploit the technological capabilities of emerging micro- and nanotechnologies. The fusion of data from large numbers of sensors as well as large numbers of sensor types will drive research in new networking concepts.

Recommendation T4. *The Air Force should develop balanced research strategies for not only the hardware but also the requisite software and software architectures for fixed arrays and moving swarms of multispectral, multifunctional sensors.*

Finding T5. *Emerging microtechnology offers new opportunities in propulsion and aerodynamic control, in particular in (1) distributed sensors and actuators on both macro-aerodynamic surfaces and macro-aeropropulsion units and (2) new, scalable, miniaturized and distributed aero- and space-propulsion systems.* Emerging microtechnology has achieved preliminary success in sensing and controlling the boundary layer on full-size, subsonic airfoils. New devices for controlling gas and liquid flow, fabricated using microtechnology, promise to increase the power and reliability of air-breathing, full-size propulsion units. Several new aeropropulsion and space propulsion systems, such as micro turbine engines and micro rocket engines, have been fabricated and are in the early test phase.

Recommendation T5. *The Air Force should move decisively to develop new research and development programs to bring microtechnology to both macro- and microscale propulsion and aerodynamic control systems.*

Finding T6. *The Air Force strategic nanotechnology R&D plan, as presented to the committee, is focused on hardware concepts without appropriate consideration of total systems solutions.* It is well known that over the past 15 years the commercial sector has made increasing investments in architecture and software concepts to design advanced systems. The tendency has been toward codesign of the hardware and software aspects of a system. One implication of nanotechnology is that this approach will be even more essential as device capabilities continue to expand. New algorithms, architectures, and software design methods will need to be developed and employed in concert with new nanotechnology-based hardware. Investment in this strategy will enable autonomous, intelligent, self-configuring Air Force systems. The Air Force strategic plan contains many future scenarios where such systems would be the ideal, if not the only, solution.

Recommendation T6. *The Air Force should take seriously the importance of co-system design as a critical implication of continued miniaturization and should invest in the algorithm, architecture, and software R&D that will enable the codesign of hardware and software systems. This should be undertaken along with a projection of the advances that will be made in hardware.*

Finding T7. *Integration of micro- and nanoscale processes and of different material systems will be broadly important for materials, devices, and pack-*

aging. Self-assembly and directed assembly of dissimilar elements will be necessary to maximize the functionality of many micro- and nanoscale structures, devices, and systems. Achievement of high yields and long-term reliability, comparable to those of the current integrated circuit industry, will be a major challenge.

Recommendation T7. *The Air Force should monitor progress in self- and directed-assembly research and selectively invest its R&D resources. It will be critical for the Air Force to participate in developing manufacturing processes that result in reliable systems in technology areas where the military is the dominant customer—for example, in sensors and propulsion systems.* Developments in many of these areas will be driven by the commercial sector. The Air Force must stay aware of advances and apply them to its unique needs. As an example, in sensor applications a wide range of otherwise incompatible materials and fabrication processes is likely to be necessary.

Finding T8. *Four overarching themes emerge from the advance of micro- and nanotechnologies—increased information capabilities, miniaturization, new engineered materials, and increased functionality/autonomy. These themes could have a significant military impact by enabling new systems approaches to Air Force missions.*

Recommendation T8. *The Air Force should continue to examine new systems opportunities that may emerge from the successful development of micro- and nanotechnologies and use these studies to help focus its applied research and development investments in these technologies.*

Policy

The Air Force critically depends on advanced technology to accomplish its missions. In order to maintain the nation's competitive technology advantage over the long term, the Air Force must maintain a stable, robust, and effective RDT&E program. The Air Force is currently underinvesting in this critical area and has not maintained the stability necessary for sustained progress, thereby shortchanging its future and that of the nation (P1). An important new development is that the commercial sector now overshadows the military market. This means that product development is driven by commercial, not military, requirements. The DoD cannot, however, rely solely on commercial R&D and products to satisfy its needs (P2). Micro- and nanotechnologies are going to play a significant role in future Air Force systems, as detailed in the technical sections of this report (the basis for findings and recommendations T1-T8 above). The Air Force

Research Laboratory has initiated a planning process to enhance its effectiveness in this all-important area (P3), but more needs to be done to strengthen the Air Force's internal programs and to ensure that they assimilate and leverage the results of the very extensive programs under way throughout the worldwide scientific community (P4).

Finding P1. *Both overall DoD and—even more—Air Force policies have de-emphasized R&D spending to the detriment of DoD and Air Force long-term needs. The Air Force relies heavily on the technological sophistication of its platforms, systems, and weapons. Its ability to meet its long-term objectives is critically dependent on a strong and continuing commitment to R&D.*

Recommendation P1. *The Air Force must significantly increase its R&D funding levels if it is to have a meaningful role in the development of micro- and nanotechnology and if it is to be effective in harnessing these technologies for future Air Force systems.* DoD and the Air Force have historically funded a majority share of the nation's research in information technologies. Their funding retrenchment represents a national de-emphasis on the future of this critically important war-fighting capability.

Finding P2. *The military market for many micro- and nanotechnologies (e.g., advanced computing, communications, and sensing) is small in comparison with commercial markets. Yet, the Air Force and DoD have mission-specific requirements not satisfied by the commercial market. Military-specific applications will not be supported by industry without government and Air Force investment, particularly in basic research.*

Recommendation P2. *The Air Force should concentrate its efforts in micro- and nanotechnology on basic research at the front end and on Air Force-specific applications at the back end.* Rather than competing with the commercial sector, Air Force should stay strongly connected to commercial advances and adapt them to Air-Force-specific requirements.

Finding P3. *The Air Force has recognized the importance of micro- and nanotechnologies for its future capabilities and has begun a planning process to maximize the benefits of the in-house and extramural research programs. Strong leadership will be necessary to ensure maximum benefit from the Air Force Research Laboratory research programs.* The Air Force has coupled its programs with other research programs within DoD, especially those of DARPA and DDR&E.

Recommendation P3. *It will be critical to continue the planning for micro- and nanotechnologies at the highest levels of the Air Force Research Labo-*

ratory (AFRL). *AFRL should also strengthen its external review processes to assist the leadership and to ensure that its work is well coordinated with national efforts. The Air Force should coordinate its initiatives with other federal agencies and work to build collaborative programs where appropriate.*

Finding P4. *The committee perceived a lack of consistency in the quality of current in-house Air Force programs and in the benchmarking of those programs against the large number of programs under way throughout the world.*

Recommendation P4. *Considering that micro- and nanotechnology is a new and rapidly emerging interdisciplinary field, the Air Force should critically evaluate its efforts in micro- and nanotechnology to select areas of strong potential payoff for Air Force missions and to sustain the highest-quality program.* This will require the following:

- Long-term professional participation of Air Force personnel as active partners with the external micro- and nanotechnology community.
- Strong leadership and technical evaluation at the highest levels of AFRL technical leadership, as has already begun.
- Both fundamental research and focused, interdisciplinary development efforts. Fundamental research efforts are required to sustain a cadre of scientists with a deep understanding of both micro- and nanotechnology developments and of Air Force requirements. Interdisciplinary development efforts put an essential Air-Force-specific overlay on this fundamental research and force multidisciplinary teams to confront real system- and subsystem-level problems, which is essential for bringing any technology from the laboratory bench to practical application.

SPECIFIC FINDINGS AND RECOMMENDATIONS

Specific findings and recommendations are listed in this section.

Finding 3-1. *Space electronics is vital to the Air Force mission. The unique characteristics of the exoatmospheric environment place special demands on electronics that are outside the mainstream developments of the integrated circuit industry.*

Recommendation 3-1. *The Air Force must maintain a research and development effort in radiation-hardened electronics and must evaluate the continuing developments of micro- and nanotechnology for their applicability to* *space.* Some commercial developments, such as the move to silicon-on-insulator (SOI) materials, have clear radiation hardness benefits; others, such as molecular electronics, have yet to be evaluated in this context but are likely to exacerbate the problems.

Finding 3-2. *Communication is a critical aspect of information superiority.* The continuing trend to miniaturization is evident in this area as well as in computation. Traditionally, communication has been dependent on a wider materials base than computation as a result of the need for optical interactions and for high-speed analog functions at microwave and RF frequencies. The Air Force has unique communications requirements, different from those of the commercial sector, demanding sustained effort and the overlay of scientific advances with Air Force requirements.

Recommendation 3-2. *The Air Force should maintain a strong research program in both the optical and microwave/RF regimes.* MEMS technology is having a strong impact. Nanotechnology is already leading to advances in these areas as well as in computation.

Finding 3-3. *In sensors, especially for remote sensing applications of importance to the Air Force, there are a large number of high-performance requirements and military-specific functions that are clearly beyond the scope of commercial interests.*

Recommendation 3-3. *The Air Force should support in a sustained manner the research, development, and manufacturing infrastructure for military sensor systems.*

Finding 3-4. *The resurgence of the commercial MEMS sensor industry represents an opportunity for the U.S. Air Force to harvest improvements in military-specific MEMS devices.* The maturing of the overall MEMS industry and increased MEMS activity in the telecommunications market have dramatically increased the performance of design tools and the availability of complex space-qualified MEMS devices.

Recommendation 3-4. *The Air Force should actively pursue improvements and/or adaptations of commercial efforts for military needs and be prepared for a larger investment at the 6.2 through 6.4 levels.* This recommendation is not meant to imply total reliance on commercial sources. Commercial interests may drive the market in a different direction than is needed by the military, and commercial interests may interfere with military interests in the MEMS-based community.

Finding 3-5. *The application of nanoscience to structural materials is a promising area with important implications for Air Force systems. Such materials could be used for lightweight structures, improved coatings, multifunctional structures, and micromachined structures. Military-specific structural materials applications need special attention by the Air Force.* Many of the emerging nanoscience developments will march ahead without regard to potential Air Force needs or imperatives, and some will find their way into commercial products. However, some military structural materials needs will never be addressed by commercial industry. These include stealthy structures, thermal management structures for spacecraft, lighter-weight military aircraft, and structures for reduced logistics and maintenance costs. However, if the Air Force is to exploit and implement the most compelling advances, even those from the commercial world, then aerospace system designers and manufacturers and their suppliers as well as the Air Force users must constantly be aware of those advances. There need to be mechanisms to link technical developments to military system applications throughout the entire supply chain.

Recommendation 3-5. *The AFRL should continue its strong efforts in structural materials to capitalize on research and development advances. The Air Force should invest in establishing capability on the part of contractors to supply selected military-specific products at the same time as it invests at AFRL to encourage collaboration with cutting-edge researchers and the comprehensive tracking, where possible, of research and development and steering it toward Air Force needs.*

Finding 3-6. *Nanoscience as applied to the structural materials used for MEMS components is key to the successful deployment of MEMS technology.* Unresolved issues such as stiction prevention, sidewall morphology, and durability and stability of micromechanical structures are obstacles to the deployment of reliable MEMS sensors and actuators in military systems.

Recommendation 3-6. *The Air Force should focus development resources on materials issues that currently limit MEMS deployment for the military. These include structural stability, surface durability, manufacturable fabrication processes, and packaging.*

Finding 4-1. *Lithography and pattern transfer and self-assembly are key enablers for evolving micro- and nanotechnologies.*

Recommendation 4-1. *The AFRL R&D program will require access to micro- and nanolithography and pattern transfer tools. This should be accomplished using available national facilities or otherwise providing the* function internally. *Research into new nanolithography and patterning technologies, complementary to the industry push for high-throughput tools, would be a worthwhile investment.* The Air Force should not compete with industry efforts, particularly in silicon technology, but should concentrate on developing processes for structures and materials that are outside traditional silicon processing—for example, deep etching for MEMS and integration of new materials with silicon.

Finding 4-2. *So much is already known about progress in silicon, with its already highly developed and constantly improving manufacturing processes, that it is unlikely a sui generis technology will spring up sufficiently developed and robust that it will immediately supplant not only the transistor but also all the rest of the integrated circuit.* Integrated circuit technology has become extremely sophisticated, and the industry is devoting extensive resources to extending this sophistication in its drive to validate Moore's law for future IC generations. In contrast, nanotechnology is at a much earlier development stage, concentrating on the behavior of individual devices and circuit components (switches, wires, etc.). It is most likely that these new technologies will first find use as complements to silicon, not as immediate replacements for integrated circuits. Over the longer term, it is not possible to predict the relative roles of integrated circuits and new and evolving nanotechnologies.

Recommendation 4-2. *The Air Force should emphasize those areas of micro- and nanotechnology for information processing that are potentially integrable with silicon technology and that address Air Force-specific, noncommercial military applications.*

Finding 4-3. *The path from laboratory demonstration to the manufacture of reliable devices and systems is long and arduous, requiring extensive resources and prodigious technology development.* This will undoubtedly be as true of today's emerging technologies as it has been throughout the history of technology. It is worthwhile to consider this lesson when listening to the siren songs appearing daily, particularly in the popular and business press, on the future benefits of nanotechnology. There is undoubtedly an exciting future, and just as undoubtedly, we will find many surprises, both positive and negative, along the way.

Recommendation 4-3. *Air Force research efforts should be directed not only to the science of micro- and nanotechnology, but also to the development of devices and systems and wide access to the manufacturing technology required to produce them.*

Appendixes

A

Manufacturing, Design, and Reliability

SYSTEMS FOR COMPUTER-AIDED DESIGN, MANUFACTURING, AND PROCESS PLANNING

The declining costs of computers and the increasing costs of labor are changing our manufacturing society from human-dependent to machine-dependent systems. This has significant implications for nanoscale products. Electromechanical systems are often developed for the automation of internal manufacturing processes in industry. They are applied to many tasks, from simple assembly operations to large-scale material-handling processes. Machine designs are very much application-dependent, and there is an ongoing need for new electromechanical system designs either for new product development or existing process improvement. System design for an electro-mechanical machine involves mechanical and electrical control. Because of the growing importance of computers, software design is also receiving a great deal of attention.

Today, the life cycle of a product in the semiconductor market is short (usually no more than 15 months). Manufacturing model fitting and parameter estimations cannot begin until the process is fixed. Since it takes at least 3 months to fix a process and 6 months or longer to derive a suitable model and parameters, it must take less than six months for an integrated circuit with a good correlation between yield and reliability to be produced if everything is done properly and effectively. That is, after great effort and a large investment, a company can benefit from a reliable model for no more than 6 months.

BURN-IN

Since most microelectronic components ordinarily have an "infant mortality period"—that is, a period of higher failure rates on initial operation— the reliability problem during this period becomes extremely important. One purpose for requiring burn-in—initial, in-factory operation of electronic components before shipping and product installation—is to guarantee high reliability for the final product assembly. In addition, analysis of early failures provides lessons for further component improvement.

LIFE-CYCLE APPROACH

The discipline of life-cycle engineering (LCE) uses CAD/CAM and computer-aided support systems to describe many aspects of a product, including its cost; maintenance planning; reliability prediction and apportionment; mechanical, electric, and electronic failure modes; sneak circuit analysis; maintainability allocation; accessibility and testability evaluation; test point selections; and test sequencing. An LCE description can be built by a well-informed system designer. An interactive process review by specialists in various disciplines can identify deficiencies that are outside the designer's experience.

Prototypes will need to be manufactured for life testing. If the quality is acceptable, production will proceed; if not, redesign might be needed. The popularity of a product could justify its remodeling or the modification of selling strategies. If it is found that the product is not competitive, development profiles as well as field reports will be put into a database for future reference. After it has been on the market for some time, an unprofitable product may be terminated and the production equipment either sold or used for making other items.

A flowchart for a typical procedure for product design and manufacture is shown in Figure A-1. During the design and manufacturing cycle, a potential product is qualified according to the market's needs, attainable manufacturing techniques, and potential profits. Once the engineering, marketing, and finance departments have approved a product, computer designs can be initiated, followed by computer-aided design (CAD), computer-aided process planning (CAPP), and computer-aided manufacturing (CAM). Many techniques have been proposed and applied in practice through initial computer-aided design.

APPENDIX A

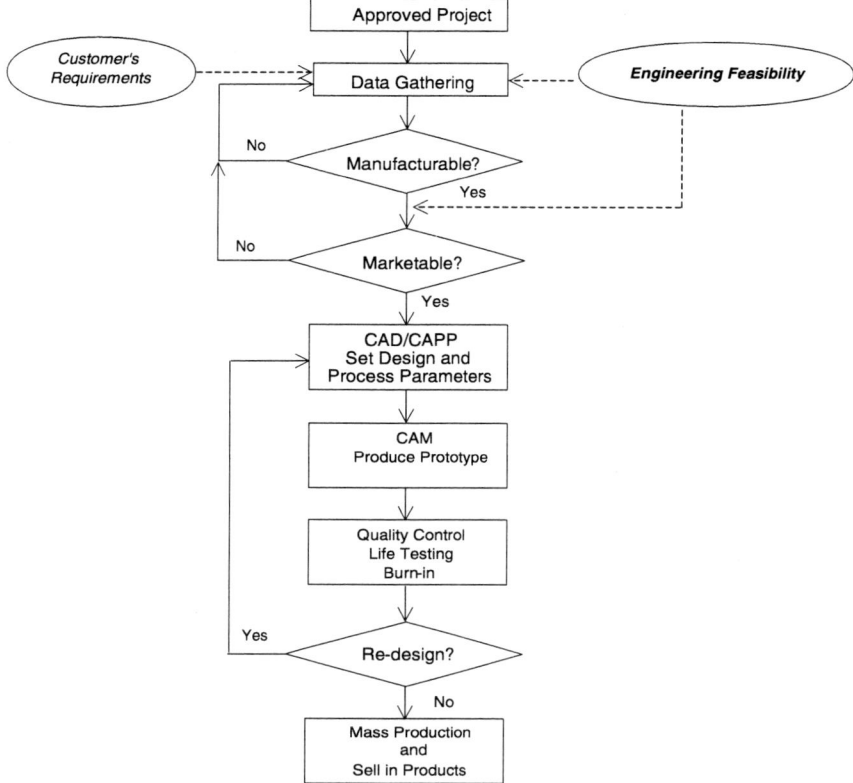

FIGURE A-1 A computerized manufacturing procedure for nanoproducts. SOURCE: Adapted from W. Kuo, W.K. Chen, and T. Kim. 1998. Reliability, Yield, and Stress Burn-In: A Unified Approach for Microelectronics Systems Manufacturing and Software Development. Norwell, Mass.: Kluwer Academic Publishers.

B

Committee Biographies

Steven R.J. Brueck *(Chair)* is the director of the Center for High Technology Materials (CHTM) and is a professor of electrical and computer engineering and a professor of physics and astronomy at the University of New Mexico. As CHTM director, he manages a research and education program that spans both optoelectronics and microelectronics. The first, optoelectronics, unites optics and electronics, and is found in CHTM's emphasis on semiconductor laser sources, optical modulators, detectors, and optical fibers. The second, microelectronics, applies semiconductor technology to the fabrication of electronic and optoelectronic devices for information and control applications. Examples of these unifying themes at work are silicon-based optoelectronics and optoelectronics for silicon manufacturing sensors. He is also a former research staff member of MIT Lincoln Laboratories. He is a member of the American Association for the Advancement of Science, the American Physical Society, the Materials Research Society, a fellow of the Institute of Electrical and Electronics Engineers, and a fellow of the Optical Society of America. His expertise includes nanoscience and nanotechnology, materials, electronics, and physics.

S. Thomas Picraux *(Vice Chair)* is the executive director of materials research, a co-director of the Center for Solid State Science in the College of Liberal Arts and Sciences, a codirector of the Center for Solid State Electronics Research in the College of Engineering and Applied Sciences, and a professor of chemical and materials engineering at Arizona State University. His current focus areas include materials that can be used to make ultrasmall devices (nanotechnology) and materials, such as wide band-gap semiconductors, that can survive in extremely harsh environments. He is a former member of the technical staff at

Sandia National Laboratories, where he also served as manager and director. At Sandia, he oversaw a variety of basic and applied science departments, including those working on nanostructure physics, optical sciences, chemical processing, biomolecular materials, surface science, and semiconductor materials and devices. He also helped develop nanotechnology research initiatives for the Department of Energy and the National Science Foundation. He is a member of the Materials Research Society and the American Vacuum Society, a senior member of the Institute of Electrical and Electronics Engineers, a fellow of the American Physical Society, and a fellow of the American Association for the Advancement of Science. He has numerous publications, including one book that he wrote and four books that he edited, and has written 12 book chapters. His expertise includes nanoscience and -technology, materials, and electronics.

John H. Belk is associate technical fellow, Phantom Works, and manager of technology planning and acquisition at the Boeing Company. He has contributed to new manufacturing processes for composite materials, optical fiber sensors for smart structures, MEMS-based sensing systems, and satellite-to-satellite laser communications while conceiving, winning, and managing millions of dollars in corporate research and development funding for Boeing in these areas. Most recently, he has been working in the Phantom Works Technology Planning and Acquisition organization to transition technologies into the Boeing Company from cutting-edge nanotechnology enterprises within venture capital funds. He received his M.S. degree in mechanical and aerospace sciences from the University of Rochester and a master's degree in engineering management from Washington University. He is now working with others across the Boeing enterprise to develop a coordinated nanoscience/nanotechnology plan for all of Boeing's activities. He holds six U.S. patents in sensing, telecommunications, and quality applications. His expertise includes process sensing.

Robert J. Celotta is a NIST fellow and leader of the electron physics group at the National Institute of Standards and Technology. His current primary research area is nanostructure science, including surface and multilayer magnetism research including polarized electron microscopy of magnetic materials; research on surface nanostructure fabrication using laser-focused deposition; nanostructure fabrication and characterization using room- and low-temperature scanning tunneling microscopy and autonomous atom assembly; research on the production and detection of electron spin polarization and the interaction of polarized electrons with atoms and solid surfaces; and research on electron collisions with atoms and molecules using monochromatic electron beams and state-selected atoms. He is a fellow of the American Physical Society, the American Vacuum Society, the American Association for the Advancement of Science, and the Washington Academy of Sciences. His expertise includes nanoscience and nanotechnology and surface physics.

William C. Holton is a visiting research professor in the department of electrical and computer engineering at North Carolina State University. His current research interest is quantum computing. He is also a former vice president for research operations at Semiconductor Research Corporation (SRC). He directed SRC's research program, participated in the definition and growth of SRC as the semiconductor industry's premier sponsor of university research, led in creation of the initial SIA Semiconductor Roadmap, and was influential in the formation of SEMATECH. Prior to joining SRC, he was director of research and development at Texas Instruments. He is a member of Phi Beta Kappa, a fellow of the Institute of Electrical and Electronics Engineers, and a fellow of the American Physical Society. He has over 120 technical publications and patent applications. His expertise includes electronics and physics.

Siegfried W. Janson is a senior scientist at the Aerospace Corporation. His current research interests are micropropulsion, microelectromechanical systems for spacecraft, and silicon satellites ("nanosats"). He invented the silicon nanosatellite and has published over 15 papers on their propulsion requirements, basic design issues, and orbital architecture. He worked in the MEMS field for over 8 years and authored or coauthored over 20 papers on microthrusters, MEMS for space applications, and silicon satellites. He flew multiple MEMS devices as a part of the MEMS testbed on the STS-93 flight and two digital microthruster arrays on a recent sounding rocket flight. He assembled a number of MEMS materials and test structures as part of the Materials on the International Space Station experiment, which was launched in late 2002. He is a member of the Institute of Electrical and Electronics Engineers and a senior member of the American Institute of Aeronautics and Astronautics. His expertise includes microscience and microtechnology.

Way Kuo, NAE, is the associate vice chancellor for engineering for the Texas A&M University system, executive associate dean for Dwight Look College of Engineering, assistant director for the Texas Engineering Experiment Station, and holder of the engineering college's Royce E. Wisenbaker Chair in Innovation. His professional interests are modeling and evaluating the reliability of modern electronic systems, with emphasis on optimal systems design. He is recognized as one of the principal scholars responsible for developing cost-effective methodologies for reducing the infant mortality period in the fast-evolving microelectronics industry. He is a fellow of the American Society for Quality, of the Institute of Industrial Engineers, and of the Institute of Electrical and Electronics Engineers. He is an elected academician of the International Academy for Quality. His expertise includes electronics.

David J. Nagel is a research professor at the George Washington University. He is now working on the development and application of MEMS and microsystems for the military and other sectors, with special attention to radio-frequency and acoustic systems. He is a former research physicist, section head, branch head, and superintendent at the Naval Research Laboratory. At NRL, his research interests centered on radiation physics, especially x-ray spectroscopy, and on materials sciences, with applications to materials analysis, plasma diagnostics, integrated circuit production, environmental studies, and MEMS. As the NRL superintendent of the condensed matter and radiation sciences division, he was a member of the Senior Executive Service and managed the experimental and theoretical research and development efforts of 150 government and contractor personnel. He is a member of the American Physical Society, the Institute for Electrical and Electronic Engineers, and the American Society of Mechanical Engineers. He has written or coauthored over 150 technical articles. He is lead author of a patent on x-ray lithography that formed the basis of a 100-person start-up company. His expertise includes microscience and microtechnology, nanoscience and nanotechnology, materials, and physics.

P. Andrew Penz is an employee and consultant with SAIC. In this position he has been developing and testing radar deinterleaving algorithms. His current work is focusing on enabling the transfer of near-real-time radar data from military platforms, e.g., the HARM Targeting System, to a national intelligence agency. Previously, he was a senior member of the technical staff in the nanoelectronic branch at Raytheon TI Systems. At Raytheon he was principal investigator on a DARPA contract to build in software an artificial nervous system (ANS) for command and control of autonomous military systems. The program produced code designed to simulate 250k high-level neurons, a computer architecture to update each neuron every millisecond, computer simulations of the dynamics of the 250k, six-layer array and a simulation showing the ability of resonant tunneling diodes to emulate axon-type active transmission line communication. He also currently holds the position of adjunct professor at the University of Texas at Dallas. He is a fellow of the American Physical Society and the Society for Information Display, a senior member of the Institute of Electrical and Electronics Engineers, and a member of the American Association for the Advancement of Science and the International Neural Network Society. His expertise includes nanoscience and nanotechnology, electronics, and physics.

Albert P. Pisano, NAE, is the director of the Electronics Research Laboratory (ERL), the director of the Berkeley Sensor and Actuator Center, a professor of mechanical engineering, and a professor of electrical engineering and computer science at the University of California at Berkeley. As director of the ERL, he manages an annual budget of about $50 million, a staff of 140 persons, and over 600 graduate students. He is in charge of day-to-day operations, long-range planning, and campus-level fundraising. His primary research interests include invention, design, fabrication, modeling, and optimization of MEMS: micro power generation devices, micro and nano resonators for RF communications, micro fluidic systems for drug delivery, micro inertial instruments, and micro information storage systems. He is a former program manager for microelectromechanical systems at the Defense Advanced Research Projects Agency (DARPA). He is a member of Pi Tau Sigma, Tau Beta Pi, and the American Society of Mechanical Engineers and an associate member of the Institute of Electrical and Electronics Engineers. He has authored or coauthored over 93 refereed publications and has graduated 20 Ph.D. students and more than 40 Masters students from the University of California at Berkeley. He frequently serves as a consultant to industry managers, academic administrators, engineering society managers, and government policy makers on MEMS research, design, application, and commercialization. His expertise includes microscience and microtechnology and nanoscience and nanotechnology.

Rosemary L. Smith is codirector of the Micro Instruments and Systems Laboratory (MISL) and a professor of electrical engineering at the University of California at Davis. Her research encompasses the design, fabrication, and applications of microfabricated sensors and instruments, including research in new microfabrication technologies and materials. She founded MISL to enhance interaction with, and serve as a microtechnology resource for, researchers in the medical and biological communities at UC Davis. MISL now has participating faculty members from dermatology, neuroscience, plant pathology, ophthalmology chemistry, veterinary medicine, molecular and cellular biology, robotics, mechanical engineering, biomedical engineering, and aeronautical engineering. She is a member of Tau Beta Pi, Eta Kappa Nu, and the Institute of Electrical and Electronics Engineers. Her expertise includes microscience and microtechnology, materials, electronics, and bioengineering.

Peter J. Stang, NAS, is dean of the College of Science and a professor of chemistry at the University of Utah. He is a preeminent physical-organic chemist and a leading experimentalist whose work has contributed significantly to modern organic chemistry. He discovered and imaginatively developed vinyl trifluoromethanesulfonates (triflates) and their chemistry. His pioneering work on the assembly of metallocyclic polygons and polyhedra is leading to new pathways to nanoscale devices and new materials. His current research interests range from physical-organic to supramolecular chemistry and self-assembly. He is a member of the American Chemical Society. He is author or coauthor of 350 publications in international and national peer-reviewed journals. His expertise includes chemistry.

George W. Sutton, NAE, is a principal engineer at the SPARTA Corporation supporting the Missile Defense Agency. He is a former director of the Washington office and chief scientist for Aero Thermo Technology, Inc., a former director of electro-optic research at Kaman Aerospace Corporation, and a former vice president of Jaycor and the Avco-Evevett Research Laboratory. Previously, he was scientific advisor to the Air Force. He is a member of the American Society of Mechanical Engineers and a fellow of both the American Association for the Advancement of Science and the American Institute of Aeronautics and Astronautics. He was editor in chief of the *AIAA Journal* for 30 years. He has written three books, more than 100 papers in technical journals and conference proceedings, many patents, and NRC reports. His doctorate is from the California Institute of Technology in mechanical engineering and physics, magna cum laude. His expertise is in military aerospace, including physics and aerospace engineering. He is a finalist for the 2002 National Medal for Technology.

William M. Tolles is a consultant. He is a former associate director of research for strategic planning and a superintendent of the chemistry division at the Naval Research Laboratory. He is also a former dean of research, dean of science and engineering, and professor of chemistry at the Naval Postgraduate School. He is a member of the American Chemical Society and the Materials Research Society. He has held long-term memberships in Phi Beta Kappa, Sigma Xi, the American Physical Society, and the Optical Society of America. He has over 50 publications. His expertise includes microscience and microtechnology, nanoscience and nanotechnology, and chemistry.

Robert J. Trew is the head of the electrical and computer engineering department and a professor of engineering at Virginia Polytechnic Institute and State University. He is also a former director of basic research in the office of the deputy under secretary of defense for science and technology and had management oversight responsibility for the $1.3 billion annual basic research programs of the Department of Defense. He was vice chair of the U.S. government interagency committee that planned and implemented the National Nanotechnology Initiative. He is a member of Sigma Xi, Eta Kappa Nu, Tau Beta Pi, the Materials Research Society, the American Association for the Advancement of Science, and the American Society for Engineering Education, is a fellow of the Institute of Electrical and Electronics Engineers, and serves on the Microwave Theory and Techniques Society ADCOM. He is currently editor of *IEEE Microwave Magazine*. He has over 140 publications and 14 book chapters and has given over 260 technical and programmatic presentations. His expertise includes nanoscience and nanotechnology and electronics.

Mary H. Young is director of the HRL Laboratories' Sensors and Materials Laboratory. She manages an organization with research emphasis in microelectromechanical (MEM) technologies, advanced energy storage and polymeric sensors, electro-optical sensor materials and process technologies, materials engineering, and nanoelectronics technologies. She has contributed original work in electronic transport physics in semiconductors and in the physics of infrared sensitive materials and IR devices. She is a member of Phi Beta Kappa, the American Physical Society, and the Materials Research Society. She has more than two dozen publications on semiconductor materials, infrared detectors, impurity hopping transport, neutron transmutation in semiconductors, and superlattice materials and devices. Her expertise includes microscience and microtechnology, nanoscience and nanotechnology, materials, electronics, and physics.

C

Meetings and Activities

FIRST MEETING, AUGUST 15-16, 2001

Sponsor Expectations, Air Force Science and Technology Overview
Don Daniel
Deputy Assistant Secretary, Air Force (Science, Technology and Engineering)

Implications of Emerging Micro and Nano-Technologies
Robert Leheny
Director, Defense Advanced Research Projects Agency

National Nanotechnology Initiative
M.C. Roco
Senior Advisor, Nanotechnology, National Science Foundation
Chair, Subcommittee on Nanoscience, Engineering and Technology

Nanoscience/Nanotechnology Programs in Office of the Secretary of Defense
Clifford Lau
Office of the Deputy Under Secretary of Defense for Science and Technology

Navy Micro and Nano Programs Overview
James Murday
Chair, Naval Nanoscience Program

Army Micro and Nano Programs Overview
Bob D. Guenther

Army Research Office, Inter-Agency Personnel Agreement, Duke University

Air Force Micro and Nano Programs Overview
Forrest Agee
Air Force Office of Scientific Research

SECOND MEETING, OCTOBER 2-3, 2001

Air Force S&T Development Planning Results: Long-Range Challenges
Gregory P. Rubertus
Chief, Corporate Investment Strategy Division, Air Force Research Laboratory (AFRL)

Overview of Fundamental Micro/Nano Research and Tool Development
Forrest Agee
Air Force Office of Scientific Research

Technical Aspects of Micro/Nano Research and Tool Development
Gernot S. Pomrenke
Air Force Office of Scientific Research

Micro/Nano Materials Development Overview
Richard Vaia
Air Force Research Laboratory, Materials and Manufacturing Directorate

The Bio/Nano Interface
Morely Stone
Air Force Research Laboratory, Materials and Manufacturing Directorate

MEMS Tribology and Reliability and Nanostructured Coatings for Space and Tribology
Jeff Zabinski
Air Force Research Laboratory, Materials and Manufacturing Directorate

Nanoelectronics and Nanomaterials for Sensors
Gail Brown
Air Force Research Laboratory, Materials and Manufacturing Directorate

Micro and Nano Opportunities
Roger T. Howe
Berkeley Sensor and Actuator Center, University of California at Berkeley

Recent Accomplishments and Future Work in MEMS
Carl Tilmann
Air Force Research Laboratory, Air Vehicles Directorate

Recent Accomplishments and Future Work in Nanotechnology
William Baron
Air Force Research Laboratory, Air Vehicles Directorate

Overview of MEMS for Propulsion and Power Applications
Kirk Yerkes
Air Force Research Laboratory, Propulsion Directorate

Overview of Nano Technology for Propulsion and Power Applications
James R. Gord
Air Force Research Laboratory, Propulsion Directorate

RF MEMS Device Technology
John L. Ebel
Air Force Research Laboratory, Sensors Directorate

RF MEMS Applications
Stephen Schneider
Air Force Research Laboratory, Sensors Directorate

Nanotechnologies for Sensors
Ross Dettmer
Air Force Research Laboratory, Sensors Directorate

Micro and Nano Opportunities
James Ellenbogen
Principal Scientist, Nanosystems Group, MITRE Corporation

THIRD MEETING, NOVEMBER 8–9, 2001

Role of Nano Technologies and Micro Systems for Space
Harald Schone and Joseph Tringe
Air Force Research Laboratory, Space Vehicles Directorate

Applications of Nano Technologies in Directed Energy
Roy Hamil and Don Shiffler
Air Force Research Laboratory, Directed Energy Directorate

Impact of Nanoenergetic Materials
Ronald Armstrong
Air Force Research Laboratory, Munitions Directorate

MEMS/Nano Technology for Assembly of Systems
John Randall
Chief Technical Officer and Vice President of Research, Zyvex

Molecules for Electronics and Bio Tech
Herb Goronkin
Vice President and Director, Physical Research Laboratories, Motorola Labs

Implications of Emerging Micro and Nano Technologies on Information Directorate Mission
Daniel Burns
Air Force Research Laboratory, Information Directorate

Micro and Nano Opportunities, Emerging Technology
George I. Bourianoff
Intel Corporation

Micro and Nano Opportunities
Terry A. Michalske
Biomolecular Materials and Interfaces Department, Sandia National Laboratories

FOURTH MEETING, DECEMBER 18–19, 2001

Nano Science and Technology Initiatives
Barbara Wilson
Air Force Research Laboratory

MEMS Technology at Draper Laboratory
Amy Duwel
Draper Laboratory

Micro/Pico Satellites
Ernest Robinson
Aerospace Corporation

Nanomechanics
Michael Roukes
California Institute of Technology

Emerging Applications of Micro and Nanophotonic Components and Their Interface with Biology
Sadik Esener
University of California, San Diego

Polymer Chemistry and Nanoscience
Grant Willson
University of Texas

Nanometals and Air Force Technology Development
Terry C. Lowe
CEO Metallicum, LLC

FIFTH MEETING, FEBRUARY 20–21, 2002

Nanotechnology for Sensors
Paul M. Amirtharaj
Sensors and Electron Devices Directorate, Army Research Laboratory

SIXTH MEETING, APRIL 2-3, 2002

Writing Session